Guido Pintacuda

NMR and NIR-CD Studies on Lanthanide Complexes in Solution

TESI DI PERFEZIONAMENTO

SCUOLA NORMALE SUPERIORE
2004

ISBN: 88-7642-147-5

"Gia. Ca. che scrive queste cose con l'amarezza nel cuore, sa benissimo che gli può capitare la disgrazia d'essere preso prima di uscire dallo Stato e di venir riportato nelle mani di quella giustizia alla cui spada oggi si sottrae. In questo caso si rivolge fin d'ora all'umanità dei suoi generosi giudici e li supplica di non voler rendere ancora più crudele la sua sorte, punendolo per quel che egli ha fatto sotto la spinta della ragione e della natura."

Giacomo Casanova
La mia fuga dai Piombi

a Guido Sacchi, che cercava di convincerci che
la scienza fosse più seria della letteratura

Aknowledgements

Che senso avrebbe una tesi se non ringraziassi tutti i miei parenti, in particolare le mie nonne Graziella ed Elena, le mie sorelle Giovanna e Greta, mia mamma, mio papà, Michi e Felice? E soprattutto i miei zii di Pavia Lino e Fiorella, senza i quali non avrei avuto il coraggio di mettermi in questo pasticcio? E se non ricordassi il cuggino Emiliano, e tutti i siciliani, con cui abbiamo tentato (e per fortuna fallito) di trasformare la Scuola Normale in un feudo casalingo? No, non avrebbe proprio nessun senso. Per non parlare delle persone che a mano a mano, negli anni, sono diventate la mia famiglia a Pisa: si può forse tacere di Riccardo, Chiara, Omar, Francesca Maria, Rino, Moreno e Silvia, Lorenzo e Emanuela, Beatrice, Chiara, Lorenzo, Cristina, Simona, Andrea, Matteo *il Re*, Giorgio, Roberta, Anna, Alessandro? E poi dimenticare Stefania, Serena, Claudia, Simonetta, Francesca, Fabrizio, Chiara, Simona, Laura o Pietro, magari solo perché sono più piccoli? Oppure Stefano (*il bacìn!*), Andrea (*il trave*)? Sarebbe come, che so, non nominare Orsetta, *lo Zà*, Matteo, Mariangela, Luca: una follia.
E come faccio poi a non parlare degli amici americani di Berkeley (yes yes, I've got American friends!): Barbara, Linda, Rob, Fran, David, Ruben and Nacho? O di quelli francesi: Matteo *le Roi*, Delphine, Marco, Fabien, Emilie, Séverine, Anne et Jean-Christophe??

Il fatto è che i ringraziamenti in una tesi sono una cosa ufficiale. Ecco che allora, *more specifically, I thank my supervisors prof. Piero Salvadori and Lorenzo di Bari for constant support and stimulating discussions; Rino Pescitelli, Moreno Lelli and Silvia Ripoli, coauthors of most of the work described in the present thesis; my father, Roberto Simeone and Fabio Merotto for the help in handling the graphic files; Angela Cuzzola and Ilaria Bienintesi, for the help in the ESI-MS spectra and in the organic synthesis; Francesca Maria Marchetti for typing most of the equations of chapter 2. I am also grateful to prof. David Parker and Rachel Dickins for providing some samples studied in chapter 3 and for running the fluorescence measurements.*

The present document, accepted for publication in 2004, is a newer (somewhat revised) version of the work I defended at the end of my PhD. I feel indebted to prof. Gottfried Otting for a critical reading of the text, as well as to Christina Schmitz, always very sharp in sorting out my mistakes.

Non mi dimentico, infine, di Roberto Fratini: il ciclo di poesie sulle terre rare che sta progettando darà presto la polvere a questa tesi sui lantanidi.

Contents

Introduction

In the present thesis we develop new strategies concerning the use of lanthanide cations as spectroscopic probes for the study of molecular structures in solution. In the last years, lanthanides have been the object of growing interest for chemists of all the areas (inorganic, physical, organic) and they have attracted attention even of physicians and biochemists [1]. Such a widespread interest for lanthanides is mainly centered in two large fields, organic synthesis, where these ions have proved active as innovative reagents [2–5] or as (even enantioselective) catalysts for a variety of reactions [6–8], and clinical practice, where the complexes of one of them, gadolinium, lend themselves as contrast agents in magnetic resonance *imaging* (MRI), which is an effective tool for the modern medical diagnosis [9]. On the other hand, also in contexts where rare earths do not exert any chemical function, they can still be profitably used as probes for structural determination purposes, for example as radioactive or fluorescent labels for medical and biological studies [10–12]; as *shift reagents* in NMR [13]; and in particular can substitute in several systems of biological interest the Ca^{2+} ion, which does not possess suitable spectroscopic features for allowing a direct observation of its binding to biomolecules [14].

In other words, systems containing lanthanides enjoy a particularly favorable situation, since these ions feature high coordination numbers, which assure rich binding chemistry, and at the same time they possess the right spectroscopic properties to monitor it. The latter are connected to the paramagnetism of these elements, that is to the presence of unpaired f electrons on the metal centers. This key quality affects both the behavior of the nuclei surrounding the ions, to which the unpaired electrons can be coupled, and the optical behavior of the metals themselves, which can experience intraconfigurational ff transitions.

The former aspect is the object of NMR investigations, which have nowadays become a widespread and sound tool in the study of rare earth adducts in the solution state. This approach consists in measuring how the magnetic properties (chemical shift and relaxation) of a set of nuclear spins are influenced by the presence of one or more electronic spins and thus derive geometrical constraints on the relative positions of a metal ion and its ligands [15].

The second aspect is more intriguing: electronic spectroscopy of trivalent ions of lanthanide elements has been a research topic for several peculiarities. First of all, intraconfigurational ff transitions are spread in a wide range of wavelengths; in particular, several lanthanides are characterized by transitions in the near-infrared (NIR) between 800 and 2000 nm [16, 17], a spectral region where most common organic groups exhibit only weak vibrational overtones. Furthermore, owing to the strong shield exerted on the f electrons by the outer shells, a simple crystal field (CF) model can provide an accurate description of the rare earth system, and in principle both position and intensities of the experimental lines can be related to the details of the electrostatic environment surrounding a given ion [18]. In fact, an enormous amount of work has been devoted to the measurement and analysis of the $4f^N$ electronic state structures of trivalent lanthanide ions in crystals, with the leading contribution of Richardson and his coworkers [19–22], while very few reports describe investigations in the solution state. Data on the optical properties of lanthanide coordination compounds in solution (where the metal ions can perform their functional role) would lend themselves to several uses in different areas of modern chemistry [23, 24]. The present thesis develops the potentialities of electronic spectroscopy of lanthanide systems in solution. To this purpose, we point out and overcome the drawbacks which have mainly hampered this kind of study so far, namely:

- a sensitivity problem, as a result of the low absorption coefficients connected to the electrically forbidden intra-shell ff processes. The detection of CD instead of absorption contributes in overcoming this first limitation: since many of the ff transitions are magnetically allowed, they can originate circular dichroism phenomena in chiral systems and the dissymmetry factor of the resulting band will be higher the lower the absorptivity, thus turning a problem into an advantage;

- an instrumental problem, since most commercial dichrographs do not allow one to venture beyond 600-800 nm, barring this way the study of several metals. The recent development of new detectors in the field of CD has helped in solving the technical problems connected with this technique, opening the way to NIR-CD spectroscopy [25, 26]. In particular, the ion which is most advantaged by this is Yb^{3+}, which features a unique set of bands around 1000 nm. The latter originates from an isolated manifold, so mixing processes can be neglected with a consequent simplification of the spectrum analysis. Furthermore, such system is often preferred to the other lanthanides for its negligible contact contribution to the paramagnetic shift in structural studies through NMR;

- a calibration problem, due to the lack of a reasonable amount of data on systems endowed with a reasonably characterized solution structure. Several examples were reported of induced CD spectra in consequence of the interaction between rare earth ions and chiral surroundings, in particular ligands with conspicuous biological activity [27]: nevertheless, owing to the small crystal fields of the ligands, electronic lines are superimposed in solution and a large amount of potentially useful infor-mation is lost. Among other problems, it turned out difficult to present any correlation between the observed spectra and the geometries of the adducts, since only a little was known about the coordination and the solution structure and dynamics as well. To start up a rationalization, it is necessary to examine well determined molecules, and therefore to couple the analysis to an NMR study.

The overall material is organized as follows:

Part I describes the physical bases of the spectroscopies employed for the study of lanthanide ions in solution:

> **chapter 1** discusses the most relevant aspects which account for the electronic structure and the optical properties of rare earths in solids and in solution. A summary is provided both on the crystal field ap-proach, which explains the position of the f energy sub-levels, and on the different models which are commonly employed to describe the radiative transitions (absorption and CD) within the f shell;

> **chapter 2** reviews the properties of magnetic nuclei due to the inter-action with a magnetic moment of an unpaired electron in the molecule (hyperfine interaction); reference is made to the different effects of a paramagnetic ion, one concerning the shift and the other concerning the relaxation properties of the affected nucleus.

Part II describes the study of the families of chiral lanthanide complese, endowed with an axial symmetry, whose members have already been analyzed in the crystal state.

> **chapter 3** presents a first set of ligands, which is composed by C_4-symmetric analogs of DOTA (1,4,7,10-tetraazacyclododecane tetraacetic acid, **1**): this molecule is known to chelate lanthanide ions with the highest affinity in solution and it has already been studied extensively by NMR and in the MRI context, where its system with Gd^{3+} has a massive use as contrast agent.

> In the molecules examined, the introduction of chiral groups in the side chains, though far from the coordination center, dictates an helical dis-position of the coordination polyhedron and the availability of a wide variety of compounds enables the development of empirical correlations to describe the chirality transfer from the outskirts to the nucleus of these complexes, which is dependent on the position of the bulkiest groups in the ligand arms and on the nature of the donating functionalities, ranging from carboxylates to amides and alcohols;

> **chapter 4** describes how an *independent systems* model for ff transitions can simulate the experimental data in the case of Yb·**2**; the main features

DOTA 1

of the NIR-CD spectrum obtained through the theoretical computation (signal intensities and signs) show a significant correspondence with the measured values;

chapter 5 presents a second set of complexes, which is composed of mixed alkaline-lanthanide heterobimetallic chelates of the binaphthoate (BINOL) ligand of formula LnM_3BINOL_3 (Ln=lanthanide, M=alkaline), which have proved excellent catalysts in a large variety of organic reactions. These molecules adopt a C_3 symmetry and the properties of the rare earth metal are modulated by both the nature of the nature of the alkaline ions and by the chiral disposition of the BINOL units. As a central cation, Yb^{3+} is chosen for the aforementioned properties, even if the study of the correspondent diamagnetic Lu^{3+} moiety turns out to be necessary.

Part III investigates the labile interaction between lanthanide probes and a set of trial molecules of interest in organic chemistry and biochemistry, exploring the possibility to use the results collected on the model systems for characterization of wider significance. Two sets of substrates are chosen: molecules offering binding sites for the naked lanthanide ions or hosting a cavity where whole chelates can be allocated. In both cases, an equilibrium regime between free and bound forms is obtained:

chapter 6 applies NIR-CD to the investigation of the chirality of acyclic diols: through the use of an Yb^{3+} chelate, we are able to extend to the NIR region a previously described method for the study of chiral 1,2-diols; furthermore, employing Yb^{3+} as free ion, we reveal through NIR-CD the formation of at least two complexes (with different stoichiometries) between the rare earth and the set of 1,2-diols and we are able to derive an empirical correlation between the circular dichroism signals exhibited by these adducts and the absolute configuration of the bound diols;

chapter 7 performs a parallel NMR, UV-CD and NIR-CD analysis of the interaction between Ca^{2+} and rare earths with an analog of the antibiotic daunorubicin. In the case of the complex between lanthanides and an anthracyclin drug, NMR can help in locating the ion inside the principal binding site, but again NIR-CD provides information on a second binding position and on the time evolution of the adducts;

chapter 8 investigates weak interactions between a stable ytterbium chelate and cyclodextrins, as a model for the binding of contrast agents to biologically relevant macromolecules.

Part IV finally presents an account of the experimental details, an overview of the molecules analyzed and the list of the abbreviations.

Bibliography

[1] KAGAN, H. B., *Chem. Rev.* **2002**, *102*, 1805–1806.

[2] ASPINALL, H. C., *Chem. Rev.* **2002**, *102*, 1807–1850.

[3] EDELMANN, F. T.; FRECKMANN, D. M. M.; SCHUMANN, H., *Chem. Rev.* **2002**, *102*, 1851–1896.

[4] ARNDT, S.; OKUDA, J., *Chem. Rev.* **2002**, *102*, 1953–1976.

[5] KOBAYASHI, S.; SUGIURA, M.; KITAGAWA, H.; LAM, W. W.-L., *Chem. Rev.* **2002**, *102*, 2227–2302.

[6] KOBAYASHI, S., *Lanthanides: Chemistry and Use in Organic Synthesis;* SPRINGER-VERLAG: BERLIN, 1999.

[7] SHIBASAKI, M.; YOSHIKAWA, N., *Chem. Rev.* **2002**, *102*, 2187–2210.

[8] INANAGA, J.; FURUNO, H.; HAYANO, T., *Chem. Rev.* **2002**, *102*, 2211–2226.

[9] CARAVAN, P.; ELLISON, J. J.; MCMURRY, T. J.; LAUFFER, R. B., *Chem. Rev.* **1999**, *99*, 2293–2352.

[10] SELVIN, P. R., *Annu. Rev. Biophys. Biomol. Struct.* **2002**, *31*, 275–302.

[11] PARKER, D.; DICKINS, R. S.; PUSCHMANN, H.; CROSSLAND, C.; HOWARD, J. A. K., *Chem. Rev.* **2002**, *102*, 1977–2010.

[12] LIU, S.; EDWARDS, D. S., *Bioconj. Chem.* **2001**, *12*, 7–34.

[13] PETERS, J. A.; HUSKENS, J.; RABER, D. J., *Progr. NMR Spectrosc.* **1996**, *28*, 283–350.

[14] EVANS, C. H., *Biochemistry of the Lanthanides;* PLENUM PRESS: NEW YORK, 1990.

[15] BERTINI, I.; LUCHINAT, C., *Solution NMR of paramagnetic molecules. Applications to metallobiomolecules and models;* ELSEVIER: AMSTERDAM, 2001.

[16] HÜFNER, S., IN *Systematics and the properties of the lanthanides* SINHA, S. P., ED.; REIDEL: DORDRECHT, 1983.

[17] HÜFNER, S., *Optical spectra of transparent rare earth compounds;* ACADEMIC PRESS: NEW YORK, 1978.

[18] GÖRLLER-WALRAND, C.; BINNEMANS, K., "RATIONALIZATION OF CRYSTAL FIELD PARAMETRIZATION", IN *Handbook on the physics and chemistry of rare earths*, VOL. 23, K. A. GSCHNEIDER, J.; EYRING, L., EDS.; NORTH-HOLLAND: AMSTERDAM, 1996.

[19] REID, M. F.; RICHARDSON, F. S., *J. Phys. Chem.* **1984**, *88*, 3579–3586.

[20] RICHARDSON, F. S.; BERRY, M. T.; REID, M. F., *Mol. Phys.* **1986**, *58*, 929–945.

[21] BERRY, M. T.; SCHWIETERS, C.; RICHARDSON, F. S., *Chem. Phys.* **1988**, *122*, 105–124.

[22] MORAN, D. M.; RICHARDSON, F. S., *Inorg. Chem.* **1992**, *31*, 813–818.

[23] BRITTAIN, H. G., *Coord. Chem. Rev.* **1983**, *48*, 243–276.

[24] TSUKUBE, H.; SHINODA, S., *Chem. Rev.* **2002**, *102*, 1805–1806.

[25] CASTIGLIONI, E., "NIR OPERATION WITH CONVENTIONAL CD SPECTROPOLARIMETERS", IN *Book of Abstracts, 6th International Conference on CD*. PISA, 1997 .

[26] CASTIGLIONI, E.; LEBON, F.; LONGHI, G.; ABBATE, S., *Enantiomer* **2002**, *7*, 161–173.

[27] SALVADORI, P.; ROSINI, C.; BERTUCCI, C., *J. Am. Chem. Soc.* **1984**, *106*, 2439–2440.

Part I

Spectroscopies on paramagnetic substances

Chapter 1

Electronic spectroscopies with lanthanides

In this chapter we will take into account the most relevant aspects which have to be considered in order to describe the electronic structure and the optical properties of rare earths in solids and in solution. The first part of such matter has been thoroughly reviewed in the volumes by Wybourne [1] and Hüfner [2], while the physical basis and the formal treatment of the mechanisms for 4f-4f transitions have been developed in the works of Richardson [3–13], Mason, Peacock and Steward [14–17], who extended the pioneering description of the same phenomena given by Judd [18] and Ofelt [19]. To all these texts we address the reader for a more detailed explanation of the following subjects.

1.1 Electronic states of lanthanides

1.1.1 General scheme

In the following analysis, we will consider the hamiltonian $\mathcal{H}_{4f}$ for a Ln^{3+} ion in a molecular complex as the sum of a term relative to the free ion ($\mathcal{H}_{4f}(free)$) and a term describing the contribution of the crystal field ($\mathcal{H}_{cf}$):

$$\mathcal{H}_{4f} = \mathcal{H}_{4f}(free) + \mathcal{H}_{cf} \qquad (1.1)$$

The first term comprises an electrostatic component ($\mathcal{H}_{4f}(SLJ)$) and the spin-orbit interaction term ($\mathcal{H}_{so}$); the second can be conveniently separated into an achiral ($\mathcal{H}_{cf}^{g}$) and chiral ($\mathcal{H}_{cf}^{u}$) component (gerade and ungerade under coordinate inversion):

$$\mathcal{H}_{4f} = \mathcal{H}_{4f}(SLJ) + \mathcal{H}_{so} + \mathcal{H}_{cf}^{g} + \mathcal{H}_{cf}^{u} \qquad (1.2)$$

In the following, it will be discussed how:

- angular moments are vectorially added to give terms characterized by total orbital (L) and spin (S) angular moment quantum numbers;

- $\mathcal{H}_{so}$ mixes states of different S and L quantum numbers;

- $\mathcal{H}_{cf}^{g}$ splits J levels and mixes different J levels;

- $\mathcal{H}_{cf}^{u}$ mixes odd-parity states into the $4f$-electron even-parity configurational states.

1.1.2   The free ions

Rare earths are found in solids as divalent or trivalent ions, with electronic configuration $4f^{N}5s^{2}5p^{6}$ and $4f^{N-1}5s^{2}5p^{6}$ respectively. The largely prevalent valence state is anyway the trivalent, and we will consider only this latter in the rest of the discussion.

The $4f$ electrons are not the outmost: they are *screened* with respect to external fields by the two electronic shells with larger radial extension ($5s^{2}5p^{6}$) and are only weakly perturbed by the surrounding ligands. This accounts for the "atomic" nature of their spectra and explains why rare earths can constitute probes in solids and in solution: the crystal field represents only a small perturbation to the atomic energy levels and many spectroscopic properties of the salts or of the complexes can be understood considering those of the free ion. In other words, the wavefunctions of the free ions are good zeroth-order approximations for the description of the properties of the compounds.

The energy levels of a lanthanide free ion are usually interpreted considering only the interactions between the $4f$ electrons themselves. Since all the other electronic shells are spherically symmetric, their effect on all the states of a $4f$ configuration is, to a first order, the same. In this approximation, we can write the Hamiltonian determining the $4f$ energy levels as:

$$\mathcal{H} = -\frac{\hbar^{2}}{2m}\sum_{i=1}^{N}\Delta_{i} - \sum_{i=1}^{N}\frac{Z^{*}e^{2}}{r_{i}} + \underbrace{\sum_{i<j}^{N}\frac{e^{2}}{r_{ij}}}_{\mathcal{H}_{c}} + \underbrace{\sum_{i=1}^{N}\zeta(r_{i})\mathbf{s}_{i}\cdot\mathbf{l}_{i}}_{\mathcal{H}_{so}} \qquad (1.3)$$

where $N = 1,\ldots,14$ is the number of the $4f$ electrons, $Z^{*}e$ is the screened nuclear charge and $\zeta(r_{i})$ is the spin-orbit coupling function:

$$\zeta(r_{i}) = \frac{\hbar^{2}}{2m^{2}c^{2}r_{i}}\frac{dU(r_{i})}{dr_{i}} \qquad (1.4)$$

where $U(r_{i})$ is the potential to which the i-th electron is subject.

The first two terms in equation 1.3 represent the kinetic energy of the $4f$ electrons and their coulombic interaction with the nucleus: they are spherically symmetric terms and consequently they do not remove any degeneracy inside the $4f$ electron configuration. Their contribution can therefore be neglected in the rest of the discussion.

The last two terms in equation 1.3 represent the reciprocal coulombic interaction of $4f$ electrons and their spin-orbit interaction ($\mathcal{H}_{so}$), and are responsible for the energy structure of the $4f$ electrons. In the atomic theory, two limiting cases exists according to the relative importance of these two terms:

- for $\mathcal{H}_c \gg \mathcal{H}_{so}$, the so-called **Russel-Saunders coupling** (or **LS coupling**) takes place, where the spin-orbit interaction is only a small perturbation on the electronic level structure determined by diagonalizing $\mathcal{H}_c$. In this approach, the orbital angular moments for individual electrons (and likewise the spin angular moments) are vectorially summed to give terms characterized by total orbital (L) and spin (S) angular moment quantum numbers, which combine in terms as ^{2S+1}L. The spin-orbit coupling is then applied to give $(2S + 1)$ levels for each term, characterized by different values of the total angular moment quantum number J, where
$$J = |L + S|, \dots, |L - S|.$$

- for $\mathcal{H}_{so} \gg \mathcal{H}_c$, the so-called **j-j coupling** scheme occurs, in which the orbital and spin angular momenta are coupled to give a series of states for each electron, characterized by a total angular moment quantum number j. The total angular momenta of the states of each electon are then coupled with those of the other electrons to give a set of states which can be grouped in terms with different J values.

Both these limting cases are easy to face theoretically and can be solved by means of perturbation theory. In the case of lanthanides, unfortunately, the last two terms of equation 1.3 have more or less the same weight and therefore more demanding energy level calculations are required. This case is called **intermediate coupling**. It consists in adopting the Russel-Saunders scheme as a first approximation; since L and S are no longer "good" quantum numbers, while "J" is, the j-j coupling scheme is employed to evaluate the effects of the spin-orbit coupling on the levels of the Russell-Saunders terms with the same J values.

1.1.3   Crystal field hamiltonian

When a lanthanide ion is placed in a crystal lattice or in a molecular complex, it is subject to a number of complex forces which are absent in the free ion. A very simple model to face the problem is offered by the crystal field (CF) theory, an approach originally applied in isolated cases by van Vleck (1937), Abragam and Price (1951) and others for the calculation of the magnetic properties of transition metal ions in cubic symetric fields. Elliott and Stevens (1952) then developed methods for taking in account the crystal field influence in more general cases.

This model is based on the *independent systems* assumption, that is the $4f$ electrons of the Ln^{3+} and the surrounding distributions of ligands are assumed not to overlap and to interact by essentially electrostatic forces. Owing to the strong shield exerted on the $4f$ electrons by the outer shells, such an approach often allows an accurate enough description of the system.

In the most general formulation of the theory, the crystal field acts on the $4f$ free ion wave functions as a small perturbation expressed by a double multipole expansion centered on the lanthanide and on the ligand systems:

$$\mathcal{H}_{cf} = \sum_L \sum_{l_A} \sum_{l_L} V_L(l_A l_B)$$

where the $V_L(l_A l_B)$ operator is the electrostatic interaction between the l_A-pole of the Ln^{3+} charge distribution (A) and the l_B-pole of the charge distribution of the L-th ligand.[1]

It can be interesting to divide the hamiltonian $\mathcal{H}_{cf}$ in:

$$\mathcal{H}_{cf} = \mathcal{V} + \mathcal{U} \tag{1.10}$$

where

$$\mathcal{V} = \sum_L \sum_{l_A} V_L(l_A, 0) \tag{1.11}$$

$$\mathcal{U} = \sum_L \sum_{l_A} \sum_{l_L \geq 1} V_L(l_A, l_L) \tag{1.12}$$

The operator $\mathcal{V}$ (static coupling operator) represents the electrostatic interactions between the metal ion multipoles and the net charges (monopols) of the

---

[1]That is:

$$V_L(l_A, l_B) = \sum_{m_A} \sum_{m_B} T_{m_A, m_L}^{(l_A, l_L)}(L) D_{m_A}^{(l_A)}(A) D_{m_L}^{(l_L)}(L), \tag{1.5}$$

where:

$$T_{m_A, m_L}^{(l_A, l_L)}(L) = \frac{(-1)^{l_M + m_A + m_L}}{R_L^{l_A + l_L + 1}} \left(\frac{(l_A + l_L + M_A + M_L)!(l_A + l_L - M_A - M_L)!}{(l_A + m_A)!(l_A - m_A)!(l_L + m_L)!(l_L - m_L)!} \right)^{1/2}$$

$$\times C_{-m_A - m_L}^{(l_A + l_L)}(\Theta_L, \Phi_L) \tag{1.6}$$

$$D_{m_A}^{(l_A)}(A) = -\sum_\alpha e r_\alpha^{l^A} C_{m_A}^{(l_A)}(\theta_\alpha, \phi_\alpha) \tag{1.7}$$

$$D_{m_L}^{(l_L)}(L) = \sum_\beta e Z_\beta r_\beta^{l_L} C_{m_L}^{(l_L)}(\theta_\beta, \phi_\beta) \tag{1.8}$$

The chromophoric electrons on A are labelled with the α index and have coordinates $(r_\alpha, \theta_\alpha, \phi_\alpha)$; the charged particles on the L ligands are labelled with the β index and have coordinates $(r_\beta, \theta_\beta, \phi_\beta)$ and charge $Z_\beta e$ (where e is the electron charge); the ligand L position with respect of the chromophoric center A is defined by the coordinate set (R_L, Θ_L, Φ_L). The general form of the $C_m^{(l)}(\theta, \phi)$ operator is:

$$C_m^{(l)}(\theta, \phi) = \sqrt{\frac{4\pi}{(2l+1)}} Y_{l,m}(\theta, \phi), \tag{1.9}$$

where $Y_{l,m}(\theta, \phi)$ is the l-order spherical harmonic.

ligands: $\mathcal{V}$ is then the familiar *crystal field potential*. The operator $\mathcal{U}$ (dynamic coupling operator) represents the electrostatic interactions between the metal ion multipoles (l_A) and the $(l_L \geq 1)$ multipoles of the charge distributions of the ligands.

We can further divide the interaction operator $\mathcal{H}_{cf}$ in a *chiral* $\mathcal{H}_{AB}^{(u)}$ and in a achiral $\mathcal{H}_{cf}^{(g)}$:

$$\mathcal{H}_{cf} = \mathcal{H}_{cf}^{(g)} + \mathcal{H}_{cf}^{(u)} \qquad (1.13)$$

where the indexes u and g denote the transformation properties (*gerade* and *ungerade*, respectively) and:

$$\mathcal{H}_{cf}^{(g)} = \underbrace{\sum_L \sum_{l_A=2,4,6} V_L(l_A, 0)}_{\mathcal{V}_g^0} + \underbrace{\sum_L \sum_{l_A=2,4,6} V_L(l_A, 1)}_{\mathcal{U}_g^0}$$

$$\mathcal{H}_{cf}^{(u)} = \underbrace{\sum_L \sum_{l_A=1,3,5} V_L(l_A, 0)}_{\mathcal{V}_u^0} + \underbrace{\sum_L \sum_{l_A=1,3,5} V_L(l_A, 1)}_{\mathcal{U}_u^0}$$

It must be considered that:

- $\mathcal{H}_{cf}^{(g)}$, is defined to operate only *within* the $4f$ electron manifold of states, leading to J-mixing and to crystal field splittings.

$\mathcal{V}_u^0$ represents a point-charge crystal field hamiltonian which is traditionally employed in most lanthanides calculations. It determines to the first order the energy levels of the $4f$ CF states and it can be represented as a series expansion:

$$\mathcal{V}_g^0 = \sum_i \sum_{k,q} A_q^k r_i^k P_q^k(\cos\theta_i) e^{iq\phi_i} = \sum_i \sum_{k,q} A_q^k C_{i,q}^k = \sum_{k,q} B_q^k U_q^k \quad (1.14)$$

where Legendre polynomials $(P_q^k(\cos\theta_i))$, normalized spherical harmonics $(C_{i,q}^k = r_i^k P_q^k(\cos\theta_i) e^{iq\phi_i})$ or the irriducible n-particle tensors $(U_q^k = \sum_i C_{i,q}^k)$ can be used; the summation is for $0 \leq k \leq 6$ and $|q| \leq k$ and for all the $4f$ electrons (i).

By diagonalizing $\mathcal{V}_g^0$ within the *intermediate coupling* basis $|A_m\rangle$, a set of *crystal field* functions is obtained for the $4f$ electrons:

$$|A_b\rangle = \sum_m C_{bm}|A_m\rangle = \sum_{\psi JM_J} C_b(\psi JM_J)|\psi[SL]JM_J\rangle, \qquad (1.15)$$

where the expansion coefficients C_{bm} are determined by the details of the crystal field acting on the $4f$ electrons; the interaction elements $\langle A_n|\mathcal{H}_{cf}'|A_m\rangle$ can be evaluated by means of tensorial calculus;

- $\mathcal{H}_{cf}^{(u)}$ is an interconfigurational operator effective in mixing $4f$ electron states with states of opposite parity. In centrosymmetric systems, these *odd* crystal field perturbations are provided by the fluctuating potentials created by the excitation of odd parity vibrational modes in the ligands. In non-centrosymmetric systems, the most relevant contribution to such perturbations is given by the charge distributions of the ligand fixed in their equilibrium positions. This operator will promote also the coupling between the f-f transition moments on the metal ion and dissymmetric (*ungerade*) components of electric transition dipole moments localized on ligands (see below). Globally, it is responsible for the production of the electric dipole strength in f-f transitions of Ln^{3+} ions, and its sign will determine the sign of the f-f rotational strengths.

Calculation of the crystal field coefficients

The numerical values of the complex coefficients of $\mathcal{H}_{cf}^{(g)}$, A_q^k or $B_q^k = A_q^k \langle r^k \rangle$, the *crystal field parameters* (CFP), vary according to the normalization of the spherical harmonics and are usually determined fitting the experimental transition frequencies. Anyway, many attempts have been tried in the past to get these parameters theoretically (an account is given in ref. [1,2]). Given a point-charge distribution surrounding the lanthanide ion, an expansion to the first order of the hamiltonian gives:

$$\mathcal{H}_{cf}(chg) = \sum_{k,q} B_q^k(chg) U_q^k \tag{1.16}$$

with the parameters $B_q^k(chg)$ expressed by:

$$B_q^k(chg) = g(k) \sum_L T_{q,0}^{k,0} q_L \tag{1.17}$$

$$g(k) = -e(l\|r^k\|l)(l\|C^k\|l). \tag{1.18}$$

where e is the electron charge, $l = 3$ for the $4f$ electrons of the Ln^{3+} ion, $[l] = 2l + 1$ and:

$$(l\|C^k\|l) = (-1)^l [l] \begin{pmatrix} l & k & l \\ 0 & 0 & 0 \end{pmatrix} \tag{1.19}$$

The expectation values $(l\|r^k\|l)$ of r^k for the $4f$ states have been calculated by Freeman and Watson [20].

A more complete expression for the crystal field parameters was given by Faulkner and Richardson [21], who showed that another contribution to the hamiltonian, formally to the second order, can be derived to the first order assuming an "effective" crystal field potential involving the ligand polarizabilities $\bar{\alpha}_L$:

$$\mathcal{H}'_{cf}(pol) = \sum_{k,q} B_q^k(pol) U_q^k \tag{1.20}$$

where:

$$B_q^k(pol) = -2(k+1)g(k)q_A(-1)^q \sum_L \bar{\alpha}_L R_L^{-(k+4)} C_{-q}^k(\Theta_L, \Phi_L). \qquad (1.21)$$

where q_A is the lanthanide charge, R_L the distance between lanthanide and ligand L and again $g(k)$ and and C_{-q}^k are defined by equations 1.18 and 1.9, respectively.

Since the functional form of equations 1.17 and 1.20 is identical, the entire crystal field can be expressed, within the monopole/multipole approximation, as:

$$\mathcal{H}_{cf} = \sum_{k,q} \left[B_q^k(chg) + B_q^k(pol) \right] U_q^k. \qquad (1.22)$$

and the *combined* coefficients derived in this way can be easily compared with the crystal field parameters derived phenomenologically from the observed splittings.

1.2   *ff* transitions

1.2.1   Transition dipole moments and rotational strengths

The optical properties of most complexes of trivalent lanthanide ions in the UV, visible and IR regions can be explained in terms of **4f → 4f intraconfigurational radiative transitions**.

For a set of non-oriented systems (isotropic system), the oscillator strength f_{0a} and the rotational strength R_{0a} of a specific ff transition of a Ln^{3+} ion $|A_0\rangle \to |A_a\rangle$ (neglecting polarization effects of the dielectric in which the system A is immersed) are given by:

$$f_{0a} = \frac{2\pi}{\mu_B e} \frac{\chi \sigma_{0a}}{g_0} D_{0a} \qquad (1.23)$$

$$= \frac{2\pi}{\mu_B e} \frac{\chi \sigma_{0a}}{g_0} \left(|\mathbf{P}_{0a}|^2 + |\mathbf{M}_{0a}|^2 \right) \qquad (1.24)$$

$$R_{0a} = Im(\mathbf{P}_{0a} \cdot \mathbf{M}_{0a}) \qquad (1.25)$$

where D_{0a} is the transition dipole strength, $\mathbf{P}_{0a}$ e $\mathbf{M}_{0a}$ are the electric and magnetic transition dipoles, respectively; μ_B is the Bohr magneton, e the electron charge, χ is the correction of the radiative field due to the *bulk* refractivity, σ_{0a} is the transition frequency and g_0 is the degeneracy of the starting level.

Dipole and rotational strenghts are related to the experimental quantities ϵ and $\Delta\epsilon$ by the following [22]:

$$D_{0a} = \frac{3hc \, 10^3 \, ln10}{32\pi^3 N_A} \int \frac{\epsilon}{\tilde{\nu}} d\tilde{\nu} = 9.18 \cdot 10^{-3} \int \frac{\epsilon}{\tilde{\nu}} d\tilde{\nu} \qquad (1.26)$$

$$R_{0a} = \frac{3hc\,10^3\,ln10}{8\pi^3 N_A} \int \frac{\Delta\epsilon}{\tilde{\nu}}\,d\tilde{\nu} = 0.248 \int \frac{\Delta\epsilon}{\tilde{\nu}}\,d\tilde{\nu} \qquad (1.27)$$

where:

$$\epsilon = \frac{(\epsilon_L + \epsilon_R)}{2} \qquad\qquad \Delta\epsilon = \epsilon_L - \epsilon_R \qquad (1.28)$$

and ϵ_L and ϵ_R are the molar apsorption coefficients ($M^{-1}cm^{-1}$) of the left and right circularly polarized light respectively. N_A, h and c are Avogadro's number, Planck's constant and the velocity of light, respectively and the numerical factors accomodate the use of the decadic molar extinction coefficient in dm^3 mol^{-1} cm^{-1}; substitution of the values of the universal constants into equations 1.26 and 1.27 gives the dipole and rotational strenghts in units of the Debye-Bohr magneton in numerical form.

The **dissymmetry factor** is a measure of the degree of chirality of a certain transition and is defined as:

$$g = \frac{\Delta\epsilon}{\epsilon} \qquad (1.29)$$

The quantomechanical function which contains magnitude and sign of the dissymmetry factor is the R/D ratio, where R is the rotational strength and D the dipole strength of the transition. In an isotropic sample, formed by non-oriented systems, for the $0 \to a$ transition, we get:

$$\frac{R_{0a}}{D_{0a}} = \frac{Im(\mathbf{P}_{0a} \cdot \mathbf{M}_{0a})}{|\mathbf{P}_{0a}|^2 + |\mathbf{M}_{0a}|^2}. \qquad (1.30)$$

If the dominating absorption mechanism is the electric dipole ($|\mathbf{P}_{0a}|^2 \gg |\mathbf{M}_{0a}|^2$), then 1.30 becomes:

$$\frac{R_{0a}}{D_{0a}} \simeq \frac{Im(\mathbf{P}_{0a} \cdot \mathbf{M}_{0a})}{Re(\mathbf{P}_{0a} \cdot \mathbf{P}_{0a}^*)}. \qquad (1.31)$$

Choosing $\mathbf{M}_{0a}$ as pure immaginary and $\mathbf{P}_{0a}$ as pure real (as can always be done in absence of external applied fields), equation 1.31 reduces to:

$$\frac{R_{0a}}{D_{0a}} \simeq \frac{|\mathbf{M}_{0a}|}{|\mathbf{P}_{0a}|}\,\cos\eta, \qquad (1.32)$$

where η is the angle between the transition vectors $\mathbf{P}_{0a}$ and $\mathbf{M}_{0a}$.

1.2.2   Magnetic dipole transition moments

The magnetic dipole transition moments can be easily obtained in terms of spectroscopic states entirely built within the $4f^N$ configuration of the lanthanide ion by:

$$\mathbf{M}_{0a} = (A_0|\hat{\mathbf{m}}(A)|A_a) = \sum_{\psi'J'M'_J}\sum_{\psi J M_J} C_0^*(\psi J M_J)C_a(\psi'J'M'_J)$$

$$\times (\psi[SL]JM_J|\hat{\mathbf{m}}(A)|\psi'[S'L']J'M'_J). \qquad (1.33)$$

where the C_i are the crystal field coefficients of equation 1.15 and:

$$(\psi[SL]JM_J|\hat{\mathbf{m}}(A)|\psi'[S'L']J'M'_J) = \delta(\psi,\psi')\delta(S,S')\delta(L,L')(-1)^{J-M_J}$$

$$\times \begin{pmatrix} J & 1 & J' \\ -M_J & q & M'_J \end{pmatrix} ([SL]J\|\hat{\mathbf{m}}\|[SL]J'). \quad (1.34)$$

The doubly reduced matrix element on the right of the last expression can be calculated to be:

$$([SL]J\|\hat{\mathbf{m}}\|[SL]J) = 1/2\,\mu_\beta \sqrt{\frac{2J+1}{J(J+1)}}$$

$$\times \{(g_s - 1)[S(S+1) - L(L+1)] + (g_s + 1)J(J+1)\} \quad (1.35)$$

$$([SL]J\|\hat{\mathbf{m}}\|[SL]J+1) = 1/2\,\mu_\beta\,(g_s - 1)$$

$$\times \sqrt{\frac{[(S+L+1)^2 - (J+1)^2][(J+1)^2 - (L-S)^2]}{(J+1)}} \quad (1.36)$$

$$([SL]J\|\hat{\mathbf{m}}\|[SL]J-1) = 1/2\,\mu_\beta\,(1 - g_s)$$

$$\times \sqrt{\frac{[(S+L+1)^2 - J^2][J^2 - (L-S)^2]}{J}} \quad (1.37)$$

where μ_β is Bohr's magneton and g_s is the electron gyromagnetic ratio.

1.2.3 Electric dipole transition moments

Lanthanide-ligand-radiation interactions occur throught two distinct mechanisms, described by two models called **static** and **dynamic** coupling (SC and DC, respectively), according to the ligand response to the radiation:

1. the **static coupling** model is brought about by the $\mathcal{V}_u^0$ part of $\mathcal{H}_{cf}^{(u)}$, which operates within a basis set formed only by electronic states localized on the lanthanide ion and it is assumed that the electric dipole components interact directly with the $4f$ electrons;

2. the **dynamic coupling** model is promoted by the $\mathcal{U}_u^0$ part of $\mathcal{H}_{cf}^{(u)}$, which operates within a basis set composed by $4f$ states of the lanthanide ion and by a set of electronic states localized on the ligands. In this model, the electric dipole components interact with the complex throught the electronic charge distribution localized on the ligands.

These two mechanisms make separate contributions to the $4f \to 4f$ electric dipole transition momen $\mathbf{P}_{0a}$ associated with a transition $0 \to a$, namely $\mathbf{P}_{0a}^{(s)}$ and $\mathbf{P}_{0a}^{(d)}$, whose expressions have been shown to be:

$$\mathbf{P}_{0a} = \mathbf{P}_{0a}^{(s)} + \mathbf{P}_{0a}^{(d)} = \sum_q \left(\mathbf{P}_{0a;q}^{(s)} + \mathbf{P}_{0a;q}^{(s)} \right) \quad (1.38)$$

where the subscript q denotes the qth spherical component of the transition vectors. Each of the two contributions can be expanded in summation, which, taking into account the further expansion onto the crystal field states, have the form of:

$$\mathbf{P}_{0a;q}^{(s)} = \sum_{l_A}\sum_{m_A}\sum_{\psi JM_J}\sum_{\psi' J'M_J'} C_{0m}^{\star} C_{an} \left(\mathcal{A}(l_A, m_a) Z_{mn;q}^{(s)}(l_A, m_A) \right) \quad (1.39)$$

$$\mathbf{P}_{0a;q}^{(d)} = \sum_{l_A}\sum_{m_A}\sum_{\psi JM_J}\sum_{\psi' J'M_J'} C_{0m}^{\star} C_{an} \left(\mathcal{B}(l_A, m_a, q) Z_{mn}^{(d)}(l_A, m_A) \right). \quad (1.40)$$

The $\mathcal{A}$ and $\mathcal{B}$ factors depend only on the ligand properties (coordinates, charges and polarizabilities) and are defined as:

$$\mathcal{A}(l_A, m_a) = \sum_{L} q_L T_{m_A, 0}^{(l_A, 0)}(L) \quad (1.41)$$

$$\mathcal{B}(l_A, m_a, q) = (-1)^q \sum_{L} \bar{\alpha}_L T_{m_A, -q}^{(l_A, 1)}(L) \quad (1.42)$$

where the $T_{m_A, m_L}^{(l_A, l_L)}$ are the geometrical factors defined in equation 1.6.

Static coupling

The $Z_{mn;q}^{(s)}$ and $Z_{mn}^{(d)}$ depend on the $4f$ electronic structure of the Yb^{3+} ion. Letting $|A_m\rangle = |\psi[SL]JM_J\rangle$ and $|A_n\rangle = |\psi'[S'L']J'M_J'\rangle$, the $Z_{mn;q}^{(s)}$ can be calculated according to Judd[18] as:

$$\mathbf{Z}_{mn;q}^{(s)} = e^2 \sum_{\lambda, l_A, m_A} (-1)^{m_A + q}(2\lambda+1) \begin{pmatrix} 1 & \lambda & l_A \\ q & l & m_A \end{pmatrix} (-1)^{J - M_J} \begin{pmatrix} J & \lambda & J' \\ M_J & m_A + q & M_J' \end{pmatrix}$$

$$\times (\psi J \| U^{(\lambda)} \| \psi' J') \, \Xi(l_A, \lambda) \quad (1.43)$$

where $l_A = 1, 3, 5$, $m_A = 0, \pm 1, \ldots, \pm l_A$ and $\lambda = \pm l_A$. The doubly reduced matrix elements of $\mathbf{U}^{(\lambda)}$ can be obtained from the table of Nielson and Koster [23] and the $\Xi(l_A, \lambda)$ has been defined by Judd [18]:

$$\Xi(l_A, \lambda) = 2 \sum_{\bar{n}, \bar{l}} (2l+1)(2\bar{l}+1)(-1)^{l+\bar{l}} \begin{Bmatrix} 1 & \lambda & l_A \\ l & \bar{l} & l \end{Bmatrix} \begin{pmatrix} l & 1 & \bar{l} \\ 0 & 0 & 0 \end{pmatrix}$$

$$\times \begin{pmatrix} \bar{l} & l_A & l \\ 0 & 0 & 0 \end{pmatrix} \frac{\langle 4f|r|\bar{n}\bar{l}\rangle \langle \bar{n}\bar{l}|r^{l_A}|4f\rangle}{\Delta(\bar{n}\bar{l})}. \quad (1.44)$$

The summation $\sum_{\bar{n}, \bar{l}}$ is taken over the $\bar{n}d$ e $\bar{n}g$ states and $l = 3$. The computation of $\Xi(l_A, \lambda)$ and $\Delta(\bar{n}, l)$ has been conducted by Krupke [24], who extimated the values of the integrals of the kind $\langle 4f|r^{2i+1}|5d\rangle$, $\langle 4f|r^{2i+1}|n'g\rangle$ and $\langle 4f|r^{2i}|4f\rangle$ and of the "energy denominators" $\Delta(5d)$ and $\Delta(n'g)$ for a set of lanthanide ions.

Dynamic coupling

The $Z_{mn}^{(d)}$ can be calculated according to:

$$Z_{mn}^{(d)}(l_A, m_A) = -\sqrt{7}e(-1)^{J-M_J+3} \begin{pmatrix} J & l_A & J' \\ -M_J & m_A & M_J' \end{pmatrix} \begin{pmatrix} l & \lambda & l \\ 0 & 0 & 0 \end{pmatrix}$$

$$\times (\psi J \| U_l^{(\lambda)} \| \psi' J')\langle 4f|r^\lambda|4f\rangle \tag{1.45}$$

with $l_A = 2, 4, 6$. Here again the doubly reduced matrix elements of $\mathbf{U}^{(\lambda)}$ can be obtained in ref. [23], while the integrals $\langle 4f|r^{2i}|4f\rangle$ are available in Krupke's paper [24].

1.2.4   Selection rules and class of transitions

Rules based on intermediate coupling states

The selection rules for the quantum numbers of the f electrons (S,L,J ed M_J) are reported in Table 1.1, relatively to the magnetic and electric transition dipoles.

	magnetic dipole	electric dipole
weak		
ΔS	0	0
$|\Delta L|$	0	≤ 6
strong		
$|\Delta J|$	0,1	≤ 6
	(except $0 \to 0$)	(if $J, J' = 0$ then $|\Delta J| = 2,4,6$
$|\Delta M_J|$	0 (σ-pol.)	$3m_A$ (σ-pol.)
	1 (π-pol.)	$m_A \pm 1$ (π-pol.)

Table 1.1: Selection rules governing electric and magnetic dipole transitions between intermediate coupling $|\psi[SL]JM_J)$ e $|\psi'[S'L']J'M_J')$ states.

The magnetic dipole selection rules are wholy based on the intermediate coupling states $|\psi[SL]JM_J)$ constructed within the $4f^N$ configuration of Ln^{3+}. The selection rules for the electric dipole are based on the expressions 1.43 and 1.45.

As noted in the table, selection rules based on ΔS and ΔL (which refer to matrix elements between Russell-Saunders states) are expected to be much weaker for the lanthanides (except for Ce^{3+} and Yb^{3+}), while those based on ΔJ will be of primary importance for transition matrix elements between intermediate coupling states. The validity of the selection rules for J ed M_J will depend on the relative magnitude of $H_A^0 = H_A(c) + H_A(so)$ and H_{cf}, given by equations 1.3 and 1.2, respectively, and by the energy separations between the free ion terms. These selection rules will hold only when $H_A^0 \gg H_{cf}'$

or when the term levels are very spaced so that mixing between states with different J won't occur.

Classification of transitions

Based on the changes in S, L and J, a classification of ff term-to-term transitions according to their predicted dipole strengths, rotatory strengths and dissymmetry factors in a chiral ligand environment was proposed by Richardson [25].

From the expression 1.2 of the hamiltonian for a Ln^{3+} in a molecular complex, without giving any explicit consideration to the details of the crystal field interactions, but only assuming that $\mathcal{H}_{so} \gg \mathcal{H}_{cf}^g > \mathcal{H}_{cf}^u$, it can be shown which terms are essential to produce nonvanishing magnetic and electric transition dipole moments in transition types characterized by a certain ΔS, ΔL and ΔJ (Table 1.2).

	transition properties			perturbation terms	
type	$|\Delta S|$	$|\Delta L|$	$|\Delta J|$	magnetic dipole	electric dipole
1	0	0	0, 1 (J$\neq$ 0 $\neq$ J')	$-$	$\mathcal{H}_{cf}^u$
2	0	0	1 (J or J' =0)	$-$	$\mathcal{H}_{cf}^g + \mathcal{H}_{cf}^u$
3	0	> 0	0, 1 (J$\neq$ 0 $\neq$ J')	$\mathcal{H}_{cf}^g$	$\mathcal{H}_{cf}^u$
4	0	> 0	1 (J or J' =0)	$\mathcal{H}_{cf}^g$	$\mathcal{H}_{cf}^g + \mathcal{H}_{cf}^u$
5	0	$\geq$ 0	$2 \leq |\Delta J| \leq 6$ (J or J' =0)	$\mathcal{H}_{cf}^g$	$\mathcal{H}_{cf}^u$
6	0	$\geq$ 0	2, 4 or 6 (J or J' =0)	$\mathcal{H}_{cf}^g$	$\mathcal{H}_{cf}^u$
7	0	$\geq$ 0	3 or 5 (J or J' =0)	$\mathcal{H}_{cf}^g$	$\mathcal{H}_{cf}^g + \mathcal{H}_{cf}^u$
8	0	$\geq$ 0	0 (J=J'=0)	$\mathcal{H}_{cf}^g$	$\mathcal{H}_{cf}^g + \mathcal{H}_{cf}^u$
9	> 0	$\geq$ 0	0, 1 (J$\neq$ 0 $\neq$ J')	$\mathcal{H}_{so}$	$\mathcal{H}_{so} + \mathcal{H}_{cf}^u$
10	> 0	$\geq$ 0	1 (J or J' =0)	$\mathcal{H}_{so}$	$\mathcal{H}_{so} + \mathcal{H}_{cf}^g + \mathcal{H}_{cf}^u$
11	> 0	$\geq$ 0	$2 \leq |\Delta J| \leq 6$ (J$\neq$ 0 $\neq$ J')	$\mathcal{H}_{so} + \mathcal{H}_{cf}^g$	$\mathcal{H}_{so} + \mathcal{H}_{cf}^u$
12	> 0	$\geq$ 0	2, 4 or 6 (J or J' =0)	$\mathcal{H}_{so} + \mathcal{H}_{cf}^g$	$\mathcal{H}_{so} + \mathcal{H}_{cf}^u$
13	> 0	$\geq$ 0	3 or 5 (J or J' =0)	$\mathcal{H}_{so} + \mathcal{H}_{cf}^g$	$\mathcal{H}_{so} + \mathcal{H}_{cf}^g + \mathcal{H}_{cf}^u$
14	> 0	$\geq$ 0	1 (J or J' =0)	$\mathcal{H}_{so} + \mathcal{H}_{cf}^g$	$\mathcal{H}_{so} + \mathcal{H}_{cf}^g + \mathcal{H}_{cf}^u$

Table 1.2: Types of transitions between term levels: spin-orbit and crystal field perturbation based classification.

The qualitative dependence of dipole strength, rotatory strength and dissymmetry factor on the perturbative interactions $\mathcal{H}_{so}$, $\mathcal{H}_{cf}^g$ and $\mathcal{H}_{cf}^u$ can then be evaluated for each transition type on the basis of the expressions 1.24, 1.25 and 1.30 (Table 1.3).

transition types	electric dipole strengths	rotatory strengths	dissymmetry factors
1	$(\mathcal{H}_{cf}^{u})^2$	$\mathcal{H}_{cf}^{u}$	$(\mathcal{H}_{cf}^{u})^{-1}$
2	$(\mathcal{H}_{cf}^{g}\mathcal{H}_{cf}^{u})^2$	$\mathcal{H}_{cf}^{g}\mathcal{H}_{cf}^{u}$	$(\mathcal{H}_{cf}^{g}\mathcal{H}_{cf}^{u})^{-1}$
3,5,6	$(\mathcal{H}_{cf}^{u})^2$	$\mathcal{H}_{cf}^{g}\mathcal{H}_{cf}^{u}$	$(\mathcal{H}_{cf}^{g}/\mathcal{H}_{cf}^{u})$
4,7,8	$(\mathcal{H}_{cf}^{g}\mathcal{H}_{cf}^{u})^2$	$V_{g}^{2}\mathcal{H}_{cf}^{u}$	$(\mathcal{H}_{cf}^{u})^{-1}$
9	$(\mathcal{H}_{so}\mathcal{H}_{cf}^{u})^2$	$H_{so}^{2}\mathcal{H}_{cf}^{u}$	$(\mathcal{H}_{cf}^{u})^{-1}$
10	$(\mathcal{H}_{so}\mathcal{H}_{cf}^{g}\mathcal{H}_{cf}^{u})^2$	$H_{so}^{2}\mathcal{H}_{cf}^{g}\mathcal{H}_{cf}^{u}$	$(\mathcal{H}_{cf}^{g}\mathcal{H}_{cf}^{u})^{-1}$
11,12	$(\mathcal{H}_{so}\mathcal{H}_{cf}^{u})^2$	$H_{so}^{2}\mathcal{H}_{cf}^{g}\mathcal{H}_{cf}^{u}$	$(\mathcal{H}_{cf}^{g}/\mathcal{H}_{cf}^{u})$
13,14	$(\mathcal{H}_{so}\mathcal{H}_{cf}^{g}\mathcal{H}_{cf}^{u})^2$	$H_{so}^{2}V_{g}^{2}\mathcal{H}_{cf}^{u}$	$(\mathcal{H}_{cf}^{u})^{-1}$

Table 1.3: Dependence of electric dipole strengths, rotatory strengths and dissymmetry factors on spin-orbit and crystal field perturbation terms.

Finally, the qualitative classification scheme of table 1.4 can be developed, where the relative electric dipole strengths, rotatory strengths and dissymmetry factors are estimated for each term-to-term transition type, again assuming $\mathcal{H}_{so} \gg \mathcal{H}_{cf}^{g} > \mathcal{H}_{cf}^{u}$,

		class	transition types
A)	electric dipole strength (absorption and emission intensities)	E I	1,3,5,6
		E II	9,11,12
		E III	2,4,7,8
		E IV	10,13,14
B)	rotatory strength (CD and CPL intensities)	R I	1
		R II	2,3,5,6,9
		R III	4,7,8,10,11,12
		R IV	13,14
C)	dissymmetry factor	D I	2,10
		D II	1,4,7,8,9,13,14
		D III	3,5,6,11,12

Table 1.4: Classification schemes for electric dipole strengths, rotatory strengths and dissymmetry factors. The relative dipole strengths are: $EI > EII > EIII > EIV$; the relative rotatory strenghs are: $RI > RII > RIII > RIV$; the relative dissymmetry factors are: $DI > DII > DIII$.

ion	transition			transition freq. (approx), cm^{-1}
Ce^{3+}	$^2F_{5/2}$	$\rightarrow$	$^2F_{7/2}$	2100
Pr^{3+}	$^3H_4$	$\rightarrow$	$^3H_5$	2150
Nd^{3+}	$^4I_{9/2}$	$\rightarrow$	$^4I_{11/2}$	2000
Pm^{3+}	$^5I_4$	$\rightarrow$	$^5I_5$	1600
Sm^{3+}	$^6H_{5/2}$	$\rightarrow$	$^6H_{7/2}$	1400
Eu^{3+}	$^7F_1$	$\rightarrow$	$^7F_2$	1100
Tb^{3+}	$^7F_6$	$\rightarrow$	$^7F_5$	2100
Dy^{3+}	$^6H_{15/2}$	$\rightarrow$	$^6H_{13/2}$	3500
Ho^{3+}	$^5I_8$	$\rightarrow$	$^5I_7$	5000
Er^{3+}	$^4I_{15/2}$	$\rightarrow$	$^4I_{13/2}$	6400
Tm^{3+}	$^3H_6$	$\rightarrow$	$^3H_5$	5800
Yb^{3+}	$^2F_{7/2}$	$\rightarrow$	$^2F_{5/2}$	10000

Table 1.5: Transitions belonging to the RI class.

1.3   Choice of the lanthanide ion

In figures 1.1 and 1.2, the electronic $4f$ state levels and ff absorption bands are depicted for the series of all the lanthanide aquo ions. The criteria to take into account for easily studying a transition by means of CD spectroscopy are:

1. the magnitude of the rotational strength and

2. the dissymmetry factor of the transition.

Only three transition types (1,2,9) span the (RI+RII) and (DI+DII) classes of table 1.4 and should therefore prove to be the most suitable for chiroptical studies.

- with regard to **dissymmetry factors**, the only transitions in class DI which have been accessible for long time to CD measurements are the $^7F_0 \rightarrow^5 D_1$ and $^7F_0 \rightarrow^5 D_1$ excitations in Eu^{3+} and the $^7F_0 \rightarrow^5 D_1$ emission in Eu^{3+} (red arrows in fig. 1.1); the Eu^{3+} $^7F_0 \rightarrow^5 D_1$ transition (class 2) also belong to the DI class, but its excitation frequency falls in the far-infrared region ($\simeq 350$ cm^{-1});

- with regard to rotatory strengths, all the transitions of table 1.5 (blue in figures 1.1 and 1.2) are of type 1, that is RI class and therefore can be designated as "CD sensitive" transitions. Since their excitation frequencies fall in the infrared spectral region, that is in a range a frequencies outside the 180-600 nm region covered by common CD spectrometers, they have been traditionally excluded from optical activity measurements.

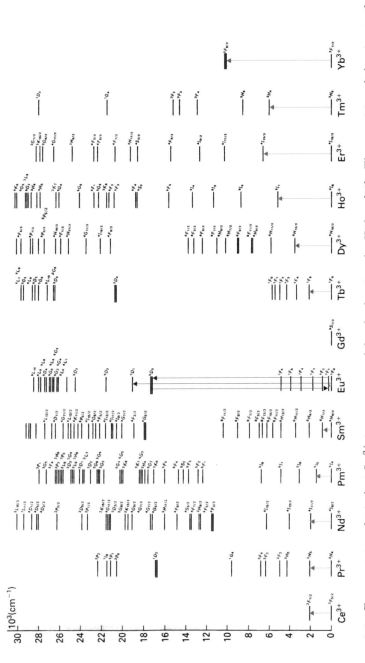

Figure 1.1: Energy states of trivalent Ln^{3+} ions as reported for the free ions by Hüfner [26]. The transitions belonging to the DI and RI classes of the Richardson classification are indicated by blue and red arrows respectively, while the levels from which fluorescent emission can occur have been colored in purple.

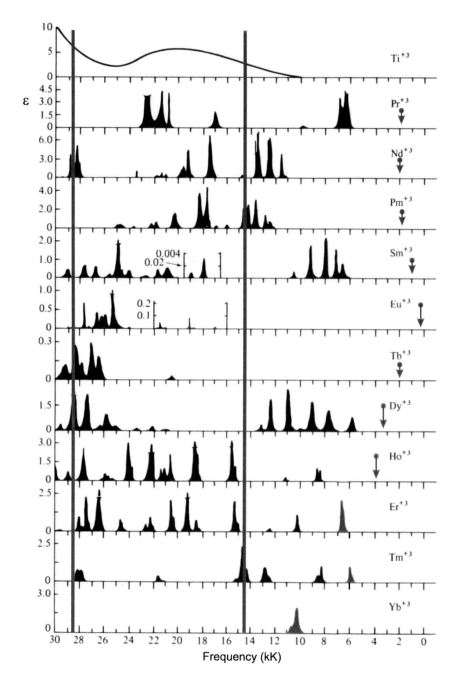

Figure 1.2: Absorption spectra of Ln^{3+} acquo ions in the Uv-Vis-IR regions (from ref. [27]). The absorption spectrum of Ti^{3+} has been added on the top for comparison.

Studies on lanthanide absorption optical activity have therefore been concerned only with RII type transitions (purple in figures 1.1 and 1.2), which are expected to originate CD bands one order of magnitude weaker than the corresponding ones in the RI class.

The developement of new instruments has now been opening the field of near-infrared CD spectroscopy; in particular, Yb^{3+} $^2F_{7/2} \rightarrow^2 F_{5/2}$ transition at about 10000 cm^{-1} is the most accessible of the whole set. The favourable magnetic properties of Yb^{3+} (see section 2.3) concur in making this ion one of the most suitable systems for structural studies in solution. Only in the last few years, and in particular during the work of this thesis, the first system-atic investigation on the CD bands originating from this transition has been performed.

1.4   Symmetric crystal fields

The number of the different crystal field sub-levels is dictated by the symmetry of the states, in turn classified according to the irreducible representations of the symmetry space group which describe their transformation properties. In this paragraph 1.4.3, an example will be given about the crystal field sub-levels in dissymmetric C_3 and C_4 ytterbium complexes, with a discussion on the possible transition types, transition rules and polarizations.

1.4.1   Symmetry of the crystal field

Let us assume G as the R_3 group of tridimensional rotations (a group with an infinite number of elements). The energy levels are characterized by their total angular moment J and the degeneration of each level is $n = 2J + 1$. Under the action of R_3, the free-ion eigenfunction transform into each other and let us assume that the representation D is irreducible. The character table of the tridimensional rotations (see [28, 29]) can be calculated by:

$$\chi_J(\alpha) = \sum_{m=-J}^{J} e^{im\alpha} = \frac{\sin\left[(2J+1)\alpha/2\right]}{\sin \alpha/2} \qquad (1.46)$$

with J integer and half-integer and α is the angle of rotation.

If a perturbation V with symmetry lower than the system hamiltoniasn $\mathcal{H}$ is introduced, only a subgroup G_v of G leaves V invariant. The D_v representation (with character $\chi_J(\alpha_v)$ of G_v can be reducible; the number of times p_s that every irreducible representation D_{vs} (with character χ_{vs}) is present in this reduction is given by:

$$p_s = \frac{1}{n} \sum_{v=1}^{c} h_v \chi_J(\alpha_v) \chi_{vs}^* \qquad (1.47)$$

where n is the order of the group (the number of symmetry operations), c is the number of classes (the number of irreducible representations D_{vs} in which

D_v can be decomposed) and h_v is the number of elements contained in the class v (the number of times a particular symmetry operation is present).
In this way, it is possible to calculate the number of crystal field states for each term J table 1.6 shows such results for all the symmetries.

symmetry		J:	0	1	2	3	4	5	6
cubic	O_h, O, T_d, T_h, T		1	1	2	3	4	4	6
hexagonal	$D_{6h}, D_6, C_{6v}, C_{6h}, C_6, D_{3h}, D_{3d},$ D_3, C_{3v}, C_3, S_6		1	2	3	5	6	7	9
tetragonal	$D_{4h}, D_4, C_{4v}, C_{4h}, D_{2d}, S_4$		1	2	4	5	7	8	10
lower symm.	$D_{2h}, D_2, C_{2v}, C_{2h}, C_2, C_s, S_2, C_1$		1	3	5	7	9	11	13

symmetry	J:	1/2	3/2	5/2	7/2	9/2	11/2	13/2
cubic		1	1	2	3	3	4	5
other symmetries		1	2	3	4	5	6	7

Table 1.6: Number of crystal field states for all the symmetries; p is the level degeneracy.

In the case of states with half-integer J, the procedure is analogous, even if the degeneracy of the levels acnnot be removen completely (Kramers theorem). As a consequence, ions with anodd number of electrons feature doubly degenerate crystal field levels, that is non-degenerate levels occur only in ions with even numbers of electrons.

1.4.2   Selection rules based on symmetry

If $D_i(A)$ and $D_j(A)$ are two representations of a group G, where A runs over the group elements, also the product:

$$D(A) = D_i(A) \times D_j(A) \tag{1.48}$$

is a representation of the group, usually reducible. This point can eb expressed as:

$$D_i(A) \times D_j(A) = \sum_l a_{ijl} D_l \tag{1.49}$$

which means that, ordering $D_i(A) \times D_j(A)$ into blocks, the irreducible representation D_l appears a_{ijl} times along the diagonal. The characters of $D_i(A) \times D_j(A)$ are given by:

$$\chi = \chi_i \cdot \chi_j. \tag{1.50}$$

These equations can be used to obtain selection rules by means of group theory. Considering a matrix element like:

$$\langle \Psi_i | \mathcal{H}_j | \Phi_l \rangle = \int \Psi_i^* \mathcal{H}_j \Phi_l d\tau \tag{1.51}$$

where Φ_i, $\mathcal{H}_j$ and Φ_l transform according to the irreducible representations D_i, D_j and D_l of their groups (let us assume real matrices, so that $D_i^* \equiv D_i$). Then the matrix element can be different from zero only if $D_i \times D_j$ contains D_l once at least. Since this holds for the characters χ_i, χ_j and χ_l, the existence of a matrix element can be obtained by the character tables (see, for example, ref. [28]).

As we saw, we are interested in the determination of the matrix elements of the electric and magnetic transition dipole moments of a certain electronic transiiton. Such matrix elements have the same symmetry properties of the cartesian coordinates (x, y, z) and of the cartesian components of the angular moment (L_x, L_y, L_z), respectively, whose characters are usually reported for each group.

1.4.3   Yb^{3+} in a trigonal or tetragonal crystal field

The Yb^{3+} ion has a $4f^{13}$ electronic configuration: it features, therefore, a single *hole* in the $4f$ shell and so a single state, ^2F. In the free ion hamiltonian, no coulombic interaction between electrons is present, but the spin-orbit coupling causes a splitting into two levels of total angular moment quantum number $J = 7/2$ and $J = 5/2$: $^2F_{7/2}$ and $^2F_{5/2}$.

Therefore, in the free ion, a single electronic transition is present, in the near infrared (NIR) region, at about 980 nm (10100 cm^{-1}) [30, 31].

In the molecules studiad in the next two chapters, the Yb^{3+} ion is subject to a crystal field of symmetries C_4 and C_3. Since the two states of Yb^{3+} have half-integer values of J, reference has to be made to the relative double groups, C_4^* and C_3^* [29, 32]. They contain 4 and 3 symmetry operations respectively (the identity and the rotations about the symmetry axis) and two irreducible representations each (E' and E'' the former, B' and E' the latter, see table 1.7). So the energy levels can be classified according to how their eigenstates transform under the symmetry operations of the double groups C_4^* and C_3^*. Each level is a Kramer doublet and can be assigned to one of the two irreducible representations E' and E'' or B' and E'.

With the information contained in table 1.8 and with formula 1.47, we can decompose the representations D_J associated to the total angular momentquantum number J; we obtain that, in both symmetries:

$$C_4 \begin{cases} D_{5/2} = E' + 2E'' \\ D_{7/2} = 2E' + 2E'' \end{cases} \qquad C_3 \begin{cases} D_{5/2} = 2B' + 2E' \\ D_{7/2} = 2B' + 3E' \end{cases}$$

So the $^2F_{5/2}$ state is splitted into 3 components and the $^2F_{7/2}$ state into 4, each doubly degenerate.

C_4	E	C_4	C_2	C_4^3	
A	1	1	1	1	z, L_z
B	1	-1	1	-1	
E {	1	i	-1	$-i$	$-i(x+iy), -(L_x+iL_y)$
	1	$-i$	-1	i	$i(x-iy), (L_x-iL_y)$

C_4^*	E	C_4	C_2	C_4^3
E' {	1	δ	i	δ^3
	1	δ^{-1}	$-i$	δ^{-3}
E'' {	1	δ^3	$-i$	δ
	1	δ^{-3}	i	δ^{-1}

C_3	E	C_3	C_3^2	
A	1	1	1	z, L_z
E {	1	ϵ	ϵ^2	$-i(x+iy), -(L_x+iL_y)$
	1	ϵ^2	ϵ	$i(x-iy), (L_x-iL_y)$

C_3^*	E	C_3	C_3^2
B'	1	-1	1
E' {	1	ω	ω^2
	1	ω^{-1}	ω^{-2}

Table 1.7: Tables of characters for the C_4, C_3, C_4^* and C_3^* groups; i is the imaginary unit, $\delta = e^{i\pi/4}$, $\epsilon = e^{2i\pi/3}$ and $\omega = e^{i\pi/3}$.

C_4	E	C_4	C_2	C_4^3
α :	0	$\pi/2$	π	$3\pi/2$
J			χ_J	
$5/2$	6	$-\sqrt{2}$	0	$\sqrt{2}$
$7/2$	8	0	0	0

C_3	E	C_3	C_3^2
α :	0	$2\pi/3$	$4\pi/3$
J		χ_J	
$5/2$	6	0	0
$7/2$	8	1	-1

Table 1.8: Characters of the C_4 and C_3 symmetry elements respect to the R_3 group; for the calculation of χ_J, formula 1.46 was employed.

The cartesian components (μ_x, μ_y) and (m_x, m_y) of the electric (μ) and magnetic (**m**) transition dipole moment operators transform as the irreducible representation E in the C_4^* and C_3^* groups and the components μ_z and m_z transform according to the irreducible representation A. Both these representations are contained in the representations of C_4^* and C_3^* of the homologous direct products $E' \times E'$ and $E'' \times E''$, while only one of them is contained in the representations of the products $B' \times B'$, $E' \times E''$ and $B' \times E'$ (see table 1.9).

C_4^*	E'^*	E''^*
E'	$2A+E$	$2B+E$
E''	$2B+E$	$2A+E$

C_3^*	B'^*	E'^*
B'	A	E
E'	E	$2A+E$

Table 1.9: Multiplication tables of the C_4^* and C_3^* double groups.

Therefore, the optical transitions between crystal field sublevels of identical symmetries $E' \to E'$, $E'' \to E''$ exhibit electric and magnetic polarization *both* parallel and perpendicular to the symmetry C_4 and C_3 axes, while the other transition kinds can only feature electric and magnetic polarization *either* perpendicular ($E' \to E''$ or $B' \to E'$) *or* parallel ($B' \to B'$) to the symmetry axis. like

Bibliography

[1] WYBOURNE, B. G., *Spectroscopic properties of rare earths*; JOHN WILEY & SONS: NEW YORK, 1965.

[2] HÜFNER, S., *Optical spectra of transparent rare earth compounds*; ACADEMIC PRESS: NEW YORK, 1978.

[3] FAULKNER, T. R.; RICHARDSON, F. S., *Mol. Phys.* **1979**, *38*, 1165–1178.

[4] RICHARDSON, F. S.; FAULKNER, T. R., *J. Chem. Phys.* **1982**, *76*, 1595–1606.

[5] SAXE, J. D.; FAULKNER, T. R.; RICHARDSON, F. S., *J. Chem. Phys.* **1982**, *76*, 1607–1623.

[6] REID, M. F.; RICHARDSON, F. S., *J. Chem. Phys.* **1983**, *79*, 5735–5742.

[7] REID, M. F.; DALLARA, J. J.; RICHARDSON, F. S., *J. Chem. Phys.* **1983**, *79*, 5743–5751.

[8] KIRKBY, A. F.; RICHARDSON, F. S., *J. Phys. Chem.* **1983**, *87*, 2544–2556.

[9] REID, M. F.; RICHARDSON, F. S., *J. Phys. Chem.* **1984**, *88*, 3579–3586.

[10] DALLARA, J. J.; REID, M. F.; RICHARDSON, F. S., *J. Phys. Chem.* **1984**, *88*, 3587–3594.

[11] RICHARDSON, F. S.; BERRY, M. T.; REID, M. F., *Mol. Phys.* **1986**, *58*, 929–945.

[12] MAY, P. S.; REID, M. F.; RICHARDSON, F. S., *Mol. Phys.* **1987**, *62*, 341–364.

[13] BERRY, M. T.; SCHWIETERS, C.; RICHARDSON, F. S., *Chem. Phys.* **1988**, *122*, 105–124.

[14] MASON, S. F.; PEACOCK, R. D.; STEWART, B., *Chem. Phys. Lett.* **1974**, *29*, 149–153.

[15] MASON, S. F.; PEACOCK, R. D.; STEWART, B., *Mol. Phys.* **1975**, *30*, 1829–1841.

[16] PEACOCK, R. D., *Struct. Bonding* **1975**, *22*, 83–109.

[17] STEWART, B., *Mol. Phys.* **1983,** *50,* 161–171.

[18] JUDD, B. R., *Phys. Rev.* **1962,** *127,* 750–761.

[19] OFELT, G. S., *J. Chem. Phys.* **1062,** *37,* 511–520.

[20] FREEMAN, A. J.; WATSON, R. E., *Phys. Rev.* **1962,** *127,* 2058–2075.

[21] FAULKNER, T. R.; RICHARDSON, F. S., *Mol. Phys.* **1980,** *39,* 75–94.

[22] MASON, S., *Molecular optical activity and the chiral discriminations;* CAMBRIDGE UNIVERSITY PRESS: CAMBRIDGE, 1982.

[23] NIELSON, C. W.; KOSTER, G. F., *Spectroscopic coefficients for p^n, d^n and f^n configurations;* MIT PRESS: CAMBRIDGE, MASSACHUSETTS, 1964.

[24] KRUPKE, W. F., *Phys. Rev.* **1966,** *145,* 325–337.

[25] RICHARDSON, F., *Inorg. Chem.* **1980,** *19,* 2806–2812.

[26] HÜFNER, S., IN *Systematics and the properties of the lanthanides* SINHA, S. P., ED.; REIDEL: DORDRECHT, 1983.

[27] CARNALL, W. T.; FIELDS, P. R., *Adv. Chem. Ser.* **1967,** *71,* 86.

[28] COTTON, F. A., *Chemical Applications of Group Theory;* WILEY: NEW YORK, 3rd ED.; 1990.

[29] BOARDMAN, A. D.; O'CONNOR, D. E.; YOUNG, P. A., *Symmetry and its Applications in Science;* MCGRAW-HILL: MAIDERHEAD, BERGSHIRE, ENGLAND, 1975.

[30] CARNALL, W. T.; FIELDS, P. R.; RAJNAK, K., *J. Chem. Phys.* **1968,** *49,* 4412–4423.

[31] CARNALL, W. T.; WYBOURNE, B. G.; FIELDS, P. R., *J. Chem. Phys.* **1965,** *42,* 3797–3805.

[32] HEINE, V., *Group Theory in Quantum Mechanics;* PERGAMON PRESS: OXFORD, 1960.

Chapter 2

NMR of paramagnetic molecules

In this section we will be concerned with the spectroscopic properties of magnetic nuclei due to the interaction with a magnetic moment of an unpaired electron in the molecule (**hyperfine interaction**) (see ref. [1]).

We will deal separately with the two main effects of a paramagnetic moiety on the magnetic properties of a given nucleus, one concerning *shift* effects and the other concerning the *relaxation properties*.

2.1 Hyperfine shift

In analyzing the coupling phenomena which influence the shift of a resonating nucleus sensing one or more unpaired electrons, it is convenient to consider the spin density **on** the nucleus and then the spin density **outside** the nucleus separately. The first part of the interaction is called **Fermi** or **contact interaction**, while the second **dipolar interaction**.

2.1.1 Fermi contact coupling

The contact shift is given by a additional magnetic field generated on the nucleus by the electron localized on the nucleus itself. Such a magnetic moment is produced by the spin density ρ on the nucleus, which, for a s orbital, is:

$$\rho = \Psi^2_{-1/2}(0) - \Psi^2_{1/2}(0) \tag{2.1}$$

where $\Psi(0)$ is the value of the MO wavefunction at zero distance from the nucleus for the -1/2 and 1/2 electron spins. The coupling between the nuclear, **I**, and the electronic, **S**, spins is proportional to the hyperfine constant $\mathcal{A}$:

$$\mathcal{A} = \frac{\mu_0}{3S} \hbar \gamma_I g_e \mu_B \sum_i \rho_i \tag{2.2}$$

where g_e is the electron g value, μ_B the Bohr magneton, γ_I the proton gyro-magnetic ratio and the sum is on all the s orbitals.

The contact-coupled spin hamiltonian in a magnetic field B_0 is given by:

$$\mathcal{H} = g_e \mu_B B_0 S_z - \hbar \gamma_I B_0 I_z + \mathcal{A} \mathbf{I} \cdot \mathbf{S} \tag{2.3}$$

$$= \frac{1}{2} Z_e - \frac{1}{2} Z_N + \mathcal{A} \left[S_z I_z + \frac{1}{2} (S_+ I_- + S_- I_+) \right] \tag{2.4}$$

where Z_e and Z_N are the electron and nuclear Zeeman energies. For $S = \frac{1}{2}$ and $I = \frac{1}{2}$, the hamiltonian matrix is:

M_S, M_I	$\lvert \frac{1}{2}, \frac{1}{2} \rangle$	$\lvert \frac{1}{2}, -\frac{1}{2} \rangle$	$\lvert -\frac{1}{2}, \frac{1}{2} \rangle$	$\lvert -\frac{1}{2}, -\frac{1}{2} \rangle$
$\langle \frac{1}{2}, \frac{1}{2} \rvert$	$\frac{1}{2} Z_e - \frac{1}{2} Z_N + \frac{1}{4} \mathcal{A}$	0	0	0
$\langle \frac{1}{2}, -\frac{1}{2} \rvert$	0	$\frac{1}{2} Z_e + \frac{1}{2} Z_N - \frac{1}{4} \mathcal{A}$	$\frac{1}{2} \mathcal{A}$	0
$\langle -\frac{1}{2}, \frac{1}{2} \rvert$	0	$\frac{1}{2} \mathcal{A}$	$-\frac{1}{2} Z_e - \frac{1}{2} Z_N - \frac{1}{4} \mathcal{A}$	0
$\langle -\frac{1}{2}, -\frac{1}{2} \rvert$	0	0	0	$-\frac{1}{2} Z_e + \frac{1}{2} Z_N + \frac{1}{4} \mathcal{A}$

From the diagonalization of such a matrix, the following energies and eigenfunctions can be obtained:

$$E_1 = \frac{1}{2} Z_e - \frac{1}{2} Z_N + \frac{1}{4} \mathcal{A} \qquad \Psi_1 = \lvert \frac{1}{2}, \frac{1}{2} \rangle$$

$$E_2 = -\frac{1}{4} \mathcal{A} + \frac{1}{2} R \qquad \Psi_2 = c_1 \lvert \frac{1}{2}, -\frac{1}{2} \rangle + c_2 \lvert -\frac{1}{2}, \frac{1}{2} \rangle$$

$$E_3 = -\frac{1}{4} \mathcal{A} - \frac{1}{2} R \qquad \Psi_3 = -c_1 \lvert \frac{1}{2}, -\frac{1}{2} \rangle + c_2 \lvert -\frac{1}{2}, \frac{1}{2} \rangle$$

$$E_4 = -\frac{1}{2} Z_e + \frac{1}{2} Z_N + \frac{1}{4} \mathcal{A} \qquad \Psi_4 = \lvert -\frac{1}{2}, -\frac{1}{2} \rangle$$

where:

$$R = \sqrt{\mathcal{A}^2 + (Z_e + Z_N)^2} \qquad c_1 = \sqrt{\frac{1}{2} \left(1 + \frac{Z_e + Z_N}{R} \right)}$$

$$c_2 = \sqrt{\frac{1}{2} \left(1 - \frac{Z_e + Z_N}{R} \right)}$$

The nuclear transition energies are thus given by $E_2 - E_1$ and $E_4 - E_3$ and, in the high-field limit $(Z_N \gg \mathcal{A})$, they are:

$$E_2 - E_1 = Z_N - \frac{1}{2}\mathcal{A} \tag{2.5}$$

$$E_4 - E_3 = Z_N + \frac{1}{2}\mathcal{A} \tag{2.6}$$

Taking into account the Boltzmann populations of the electronic levels, the average transition energy is given by:

$$\overline{\Delta E} = \frac{e^{-\frac{Z_e}{2kT}}}{e^{-\frac{Z_e}{2kT}} + e^{\frac{Z_e}{2kT}}} \left(Z_N - \frac{1}{2}\mathcal{A}\right) + \frac{e^{\frac{Z_e}{2kT}}}{e^{-\frac{Z_e}{2kT}} + e^{\frac{Z_e}{2kT}}} \left(Z_N + \frac{1}{2}\mathcal{A}\right) \tag{2.7}$$

and, in the limit $Z_e \ll 2kT$:

$$\overline{\Delta E} = \frac{1}{2}\left(1 - \frac{Z_e}{2kT}\right)\left(Z_N - \frac{1}{2}\mathcal{A}\right) + \frac{1}{2}\left(1 + \frac{Z_e}{2kT}\right)\left(Z_N + \frac{1}{2}\mathcal{A}\right) \tag{2.8}$$

$$= Z_N + \frac{1}{2}\mathcal{A}\frac{Z_e}{2kT} = \hbar\gamma_I B_0 + \mathcal{A}\frac{g_e\mu_B B_0}{4kT} \tag{2.9}$$

The contact shift (in ppm) is the ratio between the nuclear energy contribution due to coupling with the unpaired electron and the nuclear Zeeman energy, and is equal to:

$$\delta^{\mathrm{con}} = \mathcal{A}\frac{g_e\mu_B}{4\hbar\gamma_I kT} \tag{2.10}$$

that is, for systems with $S \neq \frac{1}{2}$:

$$\delta^{\mathrm{con}} = \frac{\mathcal{A}}{\hbar}\frac{g_e\mu_B S(S+1)}{3\gamma_I kT}. \tag{2.11}$$

This contribution is often written in terms of the expectation value of S_z ($\langle S_z \rangle$), according to the relation:

$$\delta^{\mathrm{con}} = \frac{\mathcal{A}}{\hbar\gamma_I B_0}\langle S_z \rangle \tag{2.12}$$

In the lanthanide ions, the orbital and spin angular momenta are strongly coupled, so that the g value is function of L, S and of their vectorial combination J:

$$g_J = 1 + \frac{J(J+1) - L(L+1) + S(S+1)}{2J(J+1)} \tag{2.13}$$

and the Zeeman term becomes $g_J\mu_B B_0 J_z$. For these systems, the *contact shift* becomes:

$$\delta^{\mathrm{con}} = \frac{\mathcal{A}}{\hbar}\frac{g_J(g_J - 1)\mu_B J(J+1)}{3\gamma_I kT}. \tag{2.14}$$

Even though this treatment doesn't take into account the crystal field separations of the J level group, it offers an useful criterium to foretell the general behaviour of the contact terms along the whole series. The $\langle S_z \rangle$ values (proportional to δ^{con} are reported in table 2.1.

2.1.2   Dipolar coupling

With *dipolar coupling* we mean a distance interaction between the magnetic dipole of a given nucleus (μ_I) and the magnetic dipole of the electron (μ_S) in the paramagnetic system. The Hamiltonian that describes interaction between these two magnetic moments is:

$$\mathcal{H} = -\frac{\mu_0}{4\pi}\left[\frac{3(\mu_I \cdot r)(\mu_S \cdot r)}{r^5} - \frac{(\mu_I \cdot \mu_S)}{r^3}\right] = -\mu_I \cdot \tilde{T} \cdot \mu_S \qquad (2.15)$$

where we introduce a dipolar tensor $\tilde{T}$:

$$\tilde{T} = \frac{1}{r^5}\begin{pmatrix} 3x^2 - r^2 & 3xy & 3xz \\ 3xy & 3y^2 - r^2 & 3yz \\ 3xz & 3yz & 3z^2 - r^2 \end{pmatrix} \qquad (2.16)$$

The magnetic moment of the proton is given by $\mu_I = \hbar\gamma_I I$ and that of the electron by $\mu_S = \hbar\gamma_S S = -g_e\mu_B S$, where I and S are proton and electron spins, γ_I and γ_S their gyromagnetic ratios, g_e the electron g factor and μ_B Bohr's magneton. The expectation value of the electron magnetic moment is related to the external magnetic field by the **magnetic susceptibility tensor** $\tilde{\chi}$:

$$\langle \mu \rangle = \tilde{\chi} \cdot \mathbf{B}_0 \qquad (2.17)$$

The dipolar hamiltonian 2.15 therefore can be rewritten as:

$$\mathcal{H} = -\hbar\gamma_I I \cdot (\tilde{T}\tilde{\chi}) \cdot \mathbf{B}_0 = \hbar\gamma_I I \cdot \tilde{\sigma} \cdot \mathbf{B}_0 \qquad (2.18)$$

where the *dipolar shift* tensor $\tilde{\sigma}$ has been introduced. Calling $\hat{k}$ the direction of the magnetic field, along which the proton spin is quantized (with spin number M_I), the orientation dependent contribution to the energy of the proton spin is:

$$E = -\int_0^{\langle \mu \rangle} \mu_I \cdot \tilde{T} \cdot d\mu_S = -\hbar\gamma_I B_0 M_I (\hat{k} \cdot \tilde{\sigma} \cdot \hat{k}) \qquad (2.19)$$

The **pseudocontact shift** is obtained averaging this term for all the possible orientations of the molecule and dividing the result for the Zeeman nuclear energy $\hbar\gamma_I B_0 M_I$; this operation is equivalent to taking the average of the energy values calculated along the principal directions $\hat{x}, \hat{y}, \hat{z}$:

$$\delta^{pc} = -\frac{1}{3}\left(\hat{x} \cdot \tilde{\sigma} \cdot \hat{x} + \hat{y} \cdot \tilde{\sigma} \cdot \hat{y} + \hat{z} \cdot \tilde{\sigma} \cdot \hat{z}\right) = -\frac{1}{3}\operatorname{Tr}(\tilde{\sigma})$$

$$= \frac{1}{12\pi r^5}\operatorname{Tr}\left[\begin{pmatrix} 3x^2 - r^2 & 3xy & 3xz \\ 3xy & 3y^2 - r^2 & 3yz \\ 3xz & 3yz & 3z^2 - r^2 \end{pmatrix}\begin{pmatrix} \chi_{xx} & \chi_{xy} & \chi_{xz} \\ \chi_{xy} & \chi_{yy} & \chi_{yz} \\ \chi_{xz} & \chi_{yz} & \chi_{zz} \end{pmatrix}\right]$$

$$(2.20)$$

In a generic reference system, the value of δ^{pc} is tehrefore:

$$\delta^{pc} = \frac{1}{12\pi\, r^5}\left[\left(\chi_{zz} - \frac{\chi_{xx} + \chi_{yy}}{2}\right)(3z^2 - r^2) + \frac{\chi_{xx} - \chi_{yy}}{2}(3x^2 - 3y^2) + \right.$$
$$\left. 6\chi_{xy}\,xy + 6\chi_{xz}\,xz + 6\chi_{yz}\,yz\right] \quad (2.21)$$

and shows a dependence on 5 parameters in the molecule-fixed coordinate system. If the axes chosen are the principal axes of $\tilde{\chi}$, the last three terms in equation 2.21 vanish, and δ^{pc} takes the form of:

$$\delta^{pc} = \frac{1}{12\pi}\left(\Delta\chi_{ax}\frac{3\cos^2\theta - 1}{r^3} + \Delta\chi_{rh}\frac{\sin^2\theta \cos 2\phi}{r^3}\right)$$
$$= \mathcal{D}_1\frac{3\cos^2\theta - 1}{r^3} + \mathcal{D}_2\frac{\sin^2\theta \cos 2\phi}{r^3} \quad (2.22)$$

where $\Delta\chi_{ax} = \chi_{zz} - (\chi_{xx} + \chi_{yy})/2$, $\Delta\chi_{rh} = (\chi_{xx} - \chi_{yy})$, $\mathcal{D}_1 = (1/12\pi)\,\Delta\chi_{ax}$, $\mathcal{D}_2 = (1/8\pi)\,\Delta\chi_{ax}$ and (r, θ, ϕ) are the polar coordinates of the nucleus under examination with respect to the electronic spin.
If the molecule is axially symmetric (C_n with $n \geq 3$), $\chi_{zz} = \chi_{\parallel}$, $\chi_{xx} = \chi_{yy} = \chi_{\perp}$ and using $\mathcal{D} = \chi_{\parallel} - \chi_{\perp}$ the formula 2.22 becomes:

$$\delta^{pc} = \frac{1}{12\pi\, r^5}\left(\chi_{\parallel} - \chi_{\perp}\right)(3z^2 - r^2) = \mathcal{D}\frac{3\cos^2\theta - 1}{r^3} \quad (2.23)$$

Figure 2.1 offers a pictorial view of the last three equations.

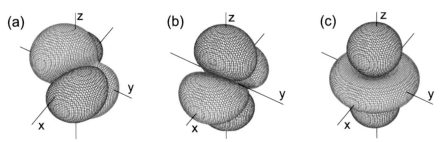

Figure 2.1: Examples of surfaces of iso-pseudocontact shifts (positive shifts in blu, negative in red) for nuclei surrounding an electronic spin. (a) In a generic reference system (equation 2.21); (b) in the principal axes system of $\tilde{\chi}$ (equation 2.22), with $\mathcal{D}_1 > 0$ and $\mathcal{D}_2 < 0$; (c) for an axially symmetric $\tilde{\chi}$ (equation 2.23), with $\mathcal{D} > 0$.

2.1.3 Bleaney's theory

The anisotropy of the magnetic susceptibility is related to the energy and shape of the electronic eigenfunctions. A model developed by Bleaney [2] relates the pseudocontact shift induced by a lanthanide ion to the crystal field parameters describing the amplitudes of the crystal field harmonics in the metal ion (chapter 1).

If the crystal field hamiltonian $\mathcal{H}_{cf}$ is expanded in a bases of spin operators (see section 1.1.3), its terms can be easily added to the electronic Zeeman hamiltonian $\mathcal{H}_z = -g_J\mu_B\mathbf{B}_0 \cdot \mathbf{J}$:

$$\mathcal{H}_t = \mathcal{H}_z + \mathcal{H}_{cf} = -g_J\mu_B\mathbf{B}_0 \cdot \mathbf{J} + \sum_{k,q} A_q^{(k)}\langle r^k\rangle\langle J||k||J\rangle O_q^{(k)} \tag{2.24}$$

where g_J is the Landè g factor, μ_B is Bohr's magneton, B_0 the magnetic field. $O_q^{(k)}$ is a spin operator which includes J_x, J_y, J_z components of total grade $k = 2, 4, 6$ with $0 \leq q \leq k$; $A_q^{(k)}$ is the crystal field coefficient, $\langle r^k\rangle$ is the average value of the k-th power of the $4f$ electron radius and $\langle J||k||J\rangle$ are numerical coefficients ($\langle J||\alpha||J\rangle$, $\langle J||\beta||J\rangle$, $\langle J||\gamma||J\rangle$ for $k = 1, 2, 3$), which can be tabulated for all the lanthanide ions (see table 2.1).

The model [2] takes the move from the expression of the principal values of the molecular magnetic susceptibility according to a generalized form of the van Vleck equation [3, 4]. In an external magnetic field $\mathbf{B}_0$ along a generic k direction, the average electronic magnetic moment $\langle\mu_{kk}\rangle$ can in fact be calculated knowing the electronic energy levels and eigenfunctions:

$$\chi_{kk} = \frac{\mu_0\langle\mu_{kk}\rangle}{B_0}$$

$$= \frac{\mu_0\mu_B^2}{kTZ}\left[\sum_A \exp\left(\frac{-E_A}{kT}\right) M_{AA}^k - 2kT\sum_{B\neq A}\frac{\exp(-E_B/kT)\,M_{AB}^k}{(E_A - E_B)}\right] \tag{2.25}$$

where μ_B is Bohr's magneton, k Boltzmann's constant and T the absolute temperature. E_A are the crystal field levels, with eigenstates $|A, i\rangle$, in which a multiplet J is split by $\mathcal{H}_{cf}$. The first sum is defined on all the crystal field sub-levels $|A, i\rangle$ within a given J term A, the second on the crystal field sub-levels $|B, i\rangle$ of the other terms B. Furthermore, $Z = \sum_{A,i}\exp(-E_A/kT)$ is the partition function, and M_{AA}^k and M_{AB}^k are related to the matrix elements of the total magnetic moment $\mathbf{M} = -g_J\mathbf{J}$:

$$M_{AA}^k = g_J^2\sum_{i,j}|\langle A, i|J^k|A, j\rangle|^2$$

$$M_{AB}^k = g_J^2\sum_{i,j}|\langle A, i|J^k|B, j\rangle|^2 \tag{2.26}$$

In a high-temperature approximation ($E_A \ll kT$), and in a situation where only the ground J multiplet is populated ($E_A - E_B \gg kT$), Bleaney proposed an expansion of χ in a power series in terms of $1/T$:

$$\chi_{kk} = \chi_{kk}^{[1]} + \chi_{kk}^{[2]} + \ldots = \sum_n \frac{C_n}{T^n} \tag{2.27}$$

Within the aforementioned conditions, the first element of the series can be calculated from equation 2.25 as:

$$\chi_{kk}^{[1]} = \frac{1}{(2J+1)\,kT} \sum_A M_{AA}^k = \frac{1}{(2J+1)\,kT} \sum_{M,M'} g_J^2 \langle J, M_J | J^k | J, M_J' \rangle |^2$$

$$= \frac{\mu_0 \mu_B^2 g_J^2 J(J+1)}{3kT}$$

$$(2.28)$$

This term is isotropic ($\chi_{xx}^{[1]} = \chi_{yy}^{[1]} = \chi_{zz}^{[1]} = \chi_{iso}$) and does not produce any pseudocontact shift (eq. 2.22). The first contribution to the pseudocontact shift arises only from the term $\chi_{kk}^{[2]}$:

$$\chi_{kk}^{[2]} = -\frac{1}{2(kT)^2(2J+1)} \sum_A g_J^2 |\langle A, i | \mathcal{H}_{cf} J_k^2 + J_k \mathcal{H}_{cf} J_k | A, j \rangle |^2 \qquad (2.29)$$

and in particular only the operators with $k = 2$ in the spin hamiltonian $\mathcal{H}_{cf}$ contribute to eq.2.29 [5,6].
The magnetic susceptibility along the x, y and z quantization axes is therefore given as:

$$\chi_{xx} - \chi_{iso} = \frac{\mu_0 \mu_B^2}{30(kT)^2} \langle r^2 \rangle (A_0^2 - A_2^2)\,\xi$$

$$\chi_{yy} - \chi_{iso} = \frac{\mu_0 \mu_B^2}{30(kT)^2} \langle r^2 \rangle (A_0^2 + A_2^2)\,\xi \qquad (2.30)$$

$$\chi_{zz} - \chi_{iso} = -\frac{\mu_0 \mu_B^2}{30(kT)^2} \langle r^2 \rangle 2A_0^2\,\xi$$

where $\langle r^2 \rangle A_0^2$ and $\langle r^2 \rangle A_2^2$ are rank-two ($k = 2$) crystal-field parameters[1] and the quantity ξ is given by:

$$\xi = g_J^2 \langle J \| \alpha \| J \rangle\, J(J+1)(2J-1)(2J+3) \qquad (2.31)$$

Higher-order crystal field parameters contribute to the power series only from the $\chi^{[3]}$ term on, having usually a minor impact in the summation [7]. The parameters determining the magnetic anisotropy of a lanthanide complex (equation 2.22) are therefore determined as:

$$\mathcal{D}_1 = \frac{\Delta \chi_{ax}}{12\pi} = -\frac{\mu_0}{4\pi} \frac{\mu_B^2}{30(kT)^2} \langle r^2 \rangle\, A_0^2\,\xi$$

$$\mathcal{D}_2 = \frac{\Delta \chi_{rh}}{8\pi} = -\frac{\mu_0}{4\pi} \frac{\mu_B^2}{30(kT)^2} \langle r^2 \rangle\, A_2^2\,\xi \qquad (2.32)$$

---

[1] They are related to the more conventional B_q^k parameters by $B_0^2 = 2\langle r^2 \rangle A_0^2$ and $B_2^2 = (2/3)^{1/2} \langle r^2 \rangle A_2^2$

ion	config.	fundam. (multipl.)	g_J	$\langle J||\alpha||J\rangle$	δ^{pc}	$-\langle S_z\rangle_J$
Ce^{3+}	$4f^1$	$^2F_{5/2}$ (6)	6/7	-0.0571	+1.6	-0.98
Pr^{3+}	$4f^2$	$^3H_4$ (9)	4/5	-0.0210	+2.7	-2.97
Nd^{3+}	$4f^3$	$^4I_{9/2}$ (6)	8/11	-0.00643	+1.05	-4.49
Pm^{3+}	$4f^4$	$^5I_4$ (9)	3/5	+0.00771	-0.6	-4.01
Sm^{3+}	$4f^5$	$^6H_{5/2}$ (6)	2/7	+0.0413	+0.17	+0.06
Eu^{3+}	$4f^6$	$^7F_0$ (1)	–	–	-1.0	+10.68
Gd^{3+}	$4f^7$	$^8S_{7/2}$ (8)	2	0	0	+31.50
Tb^{3+}	$4f^8$	$^7F_6$ (13)	3/2	-0.0101	+20.7	+31.82
Dy^{3+}	$4f^9$	$^6H_{15/2}$ (16)	4/3	-0.00635	+23.8	+28.54
Ho^{3+}	$4f^{10}$	$^5I_8$ (17)	5/4	-0.00222	+9.4	+22.63
Er^{3+}	$4f^{11}$	$^4I_{15/2}$ (16)	6/5	+0.0254	-7.7	+15.37
Tm^{3+}	$4f^{12}$	$^3H_6$ (13)	7/6	+0.0101	-12.7	+8.21
Yb^{3+}	$4f^{13}$	$^2F_{7/2}$ (8)	8/7	+0.0318	-5.2	+2.59

Table 2.1: Values of some parameters relative to lanthanide ions, from ref. [1]. In the last two columns, the expectation values of S_z ($-\langle S_z\rangle_J \propto \delta^c$) and the dipolar ($\delta^{pc}$) contributions to the nuclear resonance frequency (calculated for $r = 3$ Å, $\theta = 1$, $T = 300$ K from the D_z values exstimated by Bleaney [1, 2]) are displayed.

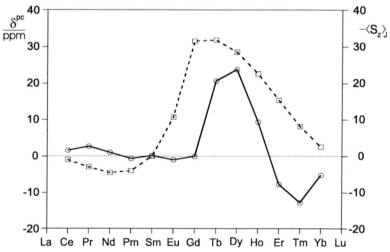

Figure 2.2: Pseudocontact (continuous line) and contact (dashed line) lanthanide induced shift in a series of homologous compounds. For the contact shifts, the expectation value of the spin magnetization $-\langle S_z\rangle_J$ is reported (table 2.1); for the pseudocontact shifts, the values are those reported in table 2.1 (see ref. [1, 2]).

If the crystal field coefficients A_0^2 and A_2^2 vary regularly in a series of isomorphic compounds containing different lanthanide ions, the *shift* is sensitive to the kind of cation (it will be greater for Tb^{3+} and Dy^{3+}) and shows a sign alternation corresponding to that of the numerical coefficient $\langle J||\alpha||J\rangle$ (see table 2.1 and figure 2.2). The theory has met several experimental verifications[8, 9], although the approximations involved may limit its applicability when the crystal field splittings exceed kT [10–12].

2.1.4 Separation of contact and pseudocontact contributions for lanthanides

Taking in account the considerations of sections 2.1.1 and 2.1.2, the NMR shift δ_{ij} induced at a nucleus i of a ligand bound to a lanthanide j can be expressed as:

$$\delta_{ij} = \delta_i^d + \delta_{ij}^c + \delta_{ij}^{pc} + \delta_j^b \qquad (2.33)$$

where δ_i^d is the diamagnetic shift (e.g. the NMR shifts of the related complexes with Ln=La, Y or Lu), δ_{ij}^c is the paramagnetic contact shift, δ_{ij}^{pc} is the paramagnetic pseudo-contact shift and δ_j^b is the shift due to the magnetic susceptibility of the sample, which can be estimated from the magnetic moment of the dissolved complex, but cancels out when all chemical shifts are referred to an internal standard.

The isotropic paramagnetic shift Δ_{ij} of a nucleus i in an axially symmetric complex containing the lanthanide j is therefore:

$$\Delta_{ij} = \delta_{ij} - \delta_j^d = \delta_{ij}^c + \delta_{ij}^{pc} = F_i \cdot \langle S_z\rangle_j + G_i \cdot \{A_0^2\langle r^2\rangle\}\frac{C_j}{T^2} \qquad (2.34)$$

- F_i is the product of the hyperfine coupling constant with a temperature dependent factor and is a function of the nucleus i and of its location in the complex;

- $G_i = (3cos^2\theta_i - 1)/r_i^3$ is the geometric factor of the nucleus i that contains the structural information about the complex (the polar coordinates r, θ of the nucleus in the principal system of the magnetic susceptibility tensor of the lanthanide):

- C_j is a magnetic constant at a given temperature T (Bleaney factor);

- $\{A_0^2\langle r^2\rangle\}$ is the crystal field parameter of eq. 1.14.

If at least two homologue paramagnetic lanthanide complexes (excluded Gd^{3+}) are available, the separation of the contact and pseudocontact contribution can be attempted, provided:

- the free-ion values of $\langle S_z\rangle_j$ are similar to those in the complexes;

- F_i is invariant for the complexes along the lanthanide transition;

- the crystal field term $\{A_0^2\langle r^2\rangle\}$ is constant along the series.

If these hypotheses are satisfied, equation 2.34 can be rearranged into equations 2.35 and 2.36:

$$\frac{\Delta_{ij}}{\langle S_z\rangle_j} = F_i + \frac{G_i}{T^2} \cdot \{A_0^2\langle r^2\rangle\} \cdot \frac{C_j}{\langle S_z\rangle_j} \tag{2.35}$$

$$\frac{\Delta_{ij}}{C_j} = F_i \cdot \frac{\langle S_z\rangle_j}{C_j} + \frac{G_i}{T^2} \cdot \{A_0^2\langle r^2\rangle\} \tag{2.36}$$

so that plots of $\frac{\Delta_{ij}}{\langle S_z\rangle_j}$ vs $\frac{C_j}{\langle S_z\rangle_j}$ or $\frac{\Delta_{ij}}{C_j}$ vs $\frac{\langle S_z\rangle_j}{C_j}$ are linear for each nucleus i within an isostructural series of complexes with lanthanide j. On the contrary, if a deviation from linearity occurs, it is usually attributed to structural changes which affect F_i and/or G_i [13].

However, it has been suggested that breaks in the plots defined by equations 2.35 and 2.36 are not necessarily associated with significant structural changes since the well-known lanthanide contraction and/or minor changes of the crystal-field parameter may induce such effects.

In order to remove the crystal field effects, Reuben [14, 15] proposed a shift modulation function which simoultaneously takes into account the shifts Δ_{ij} and Δ_{kj} of two nuclei i and k within the same ligand for various ions j.

Such crystal field independent approach has been successively extended by Geraldes et al. [16] and it has been recently applied to series of supramolecular lanthanide triple helices by Büntzli et al. [17–19]. Equation 2.34 is simultaneously solved for two different nuclei in the molecule (equations 2.37):

$$\Delta_{ij} = F_i \cdot \langle S_z\rangle_j + G_i \cdot \{A_0^2\langle r^2\rangle\} \cdot \frac{C_j}{T^2}$$
$$\Delta_{kj} = F_k \cdot \langle S_z\rangle_j + G_k \cdot \{A_0^2\langle r^2\rangle\} \cdot \frac{C_j}{T^2} \tag{2.37}$$

The pairwise ratios of such terms give equations of the form of 2.38:

$$\frac{\Delta_{ij}}{\langle S_z\rangle_j} = (F_i - F_k \cdot R_{ik}) + R_{ik} \cdot \frac{\Delta_{kj}}{\langle S_z\rangle_j} \tag{2.38}$$

with:

$$R_{ik} = \frac{G_i}{G_k} = \frac{3cos^2\theta_i - 1}{3cos^2\theta_k - 1} \cdot \frac{r_k^3}{r_i^3} \tag{2.39}$$

Plots of $\frac{\Delta_{ij}}{\langle S_z\rangle_j}$ vs $\frac{\Delta_{kj}}{\langle S_z\rangle_j}$ within an isostructural series are thus expected to be linear with a slope of R_{ik} and an intercept of $(F_i - F_k \cdot R_{ik}) + R_{ik}$. Since variations of the crystal-field parameter do not affect equation 2.38, any deviation from linearity can be safely addressed to a geometrical change along the series. Furthermore, breaks observed along the lanthanide transition according to equations 2.35 and 2.36 may also originate from variations of the hyperfine coupling constants, which can be detected from different intercepts of equation 2.38.

When contact contributions can be neglected, the pseudocontact shifts for ligand protons in a series of lanthanide complexes can be reproduced by using the complete equation for non-axial cases and the geometric factors calculated from the X-rays structures.

2.2   Relaxation

The binding of a particular lanthanide ion to a molecule causes an increase in nuclear spin-lattice relaxation rates R_1 ($= T_1^{-1}$) and spin-spin relaxation rates R_2 (T_2^{-1}) of the nearby nuclei, often referred to as lanthanide-induced relaxation enhancement (LIR) (see ref. [1], chapter 3).

There are three main different pathways by which an unpaired electron can provide fluctuating magnetic fields and cause nuclear relaxation; they are:

- **electron relaxation** between states with different M_s values, which involves changes in the orientation of the electron magnetic moment;

- **molecular rotation**, which changes the reciprocal orientation of nuclear and electron dipoles (in high-field limit both aligned with $\mathbf{B}_0$);

- **chemical exchange** of the moiety containing the resonating nucleus between a paramagnetic and a diamagnetic environment.

Each of these three phenomena is charachterized by its own correlation time, usually indicated as τ_s, electronic correlation time; τ_r, rotational correlation time; and τ_M, exchange correlation time. Electronic correlation times are usually comprised in the range $10^{-7} - 10^{-13}$ s ($10^{-8} - 10^{-9}$ s for Gd^{3+}, $10^{-12} - 10^{-13}$ s for the other lanthanides, see table 2.2); exchange correlation times can be indefinitely large or as short as 10^{-10} s; rotational correlation times can be estimated by the Stokes equation:

$$\tau_r = \frac{4\pi\eta a^3}{3kT} = \frac{\eta M_r}{dN_A kT} \tag{2.40}$$

where η is the viscosity of the solvent, a the radius of the molecule (supposed spherical), M_r the molecular weight, d its density and N_A Avogadro's constant. In water at 25°C, $\tau_r \simeq 30$ ps for aquo metal ions and it grows to $\tau_r \simeq 80$ ns for small proteins (10 KDa).

As in the case of the hyperfine shift, the nuclear relaxation induced by these contributions can be contact or dipolar in origin if spin density respectively *on* or *outside* the resonating nucleus is considered.

2.2.1   S-mechanism

It is the way by which the nucleus senses the electronic magnetic moment as fluctuating about a zero average. Nuclear relaxation originates from the modulation of the hyperfine interaction:

$$\mathcal{H}^{hf} = \mathbf{I} \cdot \tilde{\boldsymbol{T}} \cdot \mathbf{S} + \mathcal{A}\mathbf{I} \cdot \mathbf{S} = \frac{\mu_0}{4\pi}\hbar^2\gamma_I\gamma_S\left(\frac{3(\mathbf{I}\cdot\mathbf{r})(\mathbf{S}\cdot\mathbf{r})}{r^5} - \frac{\mathbf{I}\cdot\mathbf{S}}{r^3}\right) + \mathcal{A}\mathbf{I}\cdot\mathbf{S} \quad (2.41)$$

where the first term $(\mathbf{I} \cdot \tilde{\boldsymbol{T}} \cdot \mathbf{S})$ is the dipolar coupling between the electron and nuclear spin angular momenta, and the second one $(\mathcal{A}\,\mathbf{I} \cdot \mathbf{S})$ the contact coupling.

Dipolar (Solomon) contribution

It occurs when the fluctuating magnetic moment of the electron is sensed through a dipolar coupling. The energy levels and the transition frequencies ω relative to the dipolar part of the hamiltonian 2.41 are sketched in figure 2.3. Each transition is associated to a probability per unit time W: W_0 for $|\omega_I - \omega_S|$, W_1^S for ω_S, W_1^I for ω_I and W_2 for $|\omega_I + \omega_S|$.

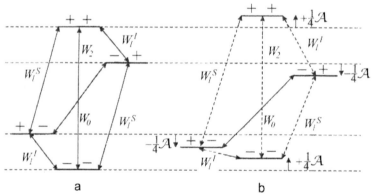

Figure 2.3: Energy level and transition frequency in a magnetically coupled two spin system. The effects of dipolar coupling (A) and contact coupling (B) are represented. W_0, W_1 and W_2 are the transition probabilities per unit time, $\mathcal{A}$ is the hyperfine coupling constant (from ref. [1]).

The nuclear longitudinal relaxation rate takes in account the probabilities of the transitions allowing nuclear spin flipping. In the case of a system of a nuclear spin I and an electron spin $S = 1/2$, its expressions gets the form (Solomon equation) [20]:

$$R_1^S = \frac{1}{10}\left(\frac{\mu_0}{4\pi}\right)^2\frac{S(S+1)\gamma_I^2 g_e^2\mu_B^2}{r^6}$$

$$\times\left(\frac{\tau_c}{1+(\omega_I-\omega_S)^2\tau_c^2} + \frac{3\tau_c}{1+\omega_I^2\tau_c^2} + \frac{6\tau_c}{1+(\omega_I+\omega_S)^2\tau_c^2}\right) \quad (2.42)$$

where the three terms in parantheses are proportional to the transition probabilities and contain the spectral densities $J(\omega, \tau_c)$ at the nuclear transition frequencies ω_I, $\omega_I + \omega_S$ and $\omega_I - \omega_S$. The correlation time is here:

$$\left(\frac{1}{\tau_c}\right)_S = \frac{1}{\tau_s} + \frac{1}{\tau_r} + \frac{1}{\tau_M} \tag{2.43}$$

Generalizing the latter result for systems with $S > \frac{1}{2}$, substituting J and g_J for S and g_e for lanthanides and approximating the values of $|\omega_I \pm \omega_S|$ by $|\omega_S|$ (the $|\omega_S|/|\omega_I|$ ratio is 658 if the nucleus I is ^1H), and working out an analogous expression for transverse relaxation [20, 21], we get:

$$R_1^S = \frac{2}{15}\left(\frac{\mu_0}{4\pi}\right)^2 \frac{J(J+1)\gamma_I^2 g_J^2 \mu_B^2}{r^6}\left(\frac{3\tau_c}{1+\omega_I^2\tau_c^2} + \frac{7\tau_c}{1+\omega_S^2\tau_c^2}\right)$$

$$R_2^S = \frac{1}{15}\left(\frac{\mu_0}{4\pi}\right)^2 \frac{J(J+1)\gamma_I^2 g_J^2 \mu_B^2}{r^6}\left(4\tau_c + \frac{3\tau_c}{1+\omega_I^2\tau_c^2} + \frac{13\tau_c}{1+\omega_S^2\tau_c^2}\right) \tag{2.44}$$

When ω_S and ω_I are much smaller than $1/\tau_c$ (fast motion limit), the equations 2.44 reduce to:

$$R_1^S = R_2^S = \frac{4}{3}\left(\frac{\mu_0}{4\pi}\right)^2 \frac{J(J+1)\gamma_I^2 g_J^2 \mu_B^2}{r^6}\tau_c \tag{2.45}$$

Ln^{3+}	$[\,H_2O\,]_6$	EDTA	DTPA
Ce	0.90	1.10	1.33
Pr	0.57	0.52	0.54
Nd	1.15	1.50	2.10
Sm	0.45	0.60	0.74
Eu	0.09		0.15
Gd		$\simeq 10^5\text{-}10^6$	
Tb	2.03	2.18	2.51
Dy	2.99	2.24	2.40
Ho	1.94	2.13	2.09
Er	2.38	1.90	1.89
Tm	3.69	1.01	2.68
Yb	1.37	0.84	1.57

Table 2.2: Calculated values of the spin relaxation times τ_s ($\cdot 10^{13}$ s) in some lanthanide(III) complexes (from ref. [22]; for the abbreviations, see appendix C).

Contact (Bloembergen) contribution

The relaxation due to the contact term is a scalar relaxation [21, 23]; the splitting of levels and the transitions relative to the contact term of hamiltonian 2.41 are depicted in figure 2.3. Rewriting this term with the formalism of spin operators, we get:

$$\mathcal{A}\mathbf{I} \cdot \mathbf{S} = \mathcal{A}(I_x S_x + I_y S_y + I_z S_z) = \frac{\mathcal{A}}{2}(S_+ I_- + S_- I_+ + 2S_z I_z) \qquad (2.46)$$

and it is evident that only flip-up and flip-down spin operators contribute to relaxation; this mechanism therefore contains only the $|\omega_I - \omega_S|$ term, while $|\omega_I + \omega_S|$ (corresponding to $S_+ I_+$, $S_- I_-$), ω_S (corresponding to $S_+ I_z$, $S_- I_z$) and ω_I (corresponding to $S_z I_+$, $S_z I_-$) do not contribute:

$$W_{\omega_I} = W_{\omega_S} = W_{(\omega_S + \omega_I)} = 0$$
$$W_{(\omega_S - \omega_I)} = \frac{1}{2}\left(\frac{\mathcal{A}}{\hbar}\right)^2 \frac{\tau_c}{1 + (\omega_S - \omega_I)^2 \tau_c^2} \qquad (2.47)$$

where the correlation time is:

$$\left(\frac{1}{\tau_c}\right)_B = \frac{1}{\tau_s} + \frac{1}{\tau_M} \qquad (2.48)$$

The contribution to the longitudinal and transverse relaxation rates (in the case $S > \frac{1}{2}$) can then be obtained as [21, 23]:

$$R_1^B = \frac{2}{3}J(J+1)\left(\frac{\mathcal{A}}{\hbar}\right)^2 \left(\frac{\tau_s}{1 + \omega_S^2 \tau_s^2}\right)$$
$$R_2^B = \frac{2}{3}J(J+1)\left(\frac{\mathcal{A}}{\hbar}\right)^2 \left(\tau_s + \frac{\tau_s}{1 + \omega_S^2 \tau_s^2}\right) \qquad (2.49)$$

where the abssence of chemical exchanges has been assumed. In the limit of $\omega_S^2 \tau_s^2 \ll 1$, the two expressions of equation 2.49 reduce to:

$$R_1^B = R_2^B = \frac{2}{3}J(J+1)\left(\frac{\mathcal{A}}{\hbar}\right)^2 \tau_s. \qquad (2.50)$$

2.2.2   Curie or χ-mechanism

It occurs when, for $\tau_e \ll \tau_r$, the electronic magnetic moment fluctuates about a non-vanishing average value, since the unequal populations of the electronic levels in thermal equilibrium produce an average magnetic moment and an average magnetic field.

The interaction of the nuclear spins with this static moment provides a further relaxation contribution, which is not modulated by electron relaxation, since the resulting electronic magnetic moment is already an average over the electron spin states.

The hamiltonian responsible for this mechanism (*susceptibility field*) is the one already described by equation 2.18:

$$\mathcal{H} = \hbar \gamma_I \mathbf{I} \cdot \tilde{\sigma} \cdot \mathbf{B}_0 \qquad (2.51)$$

The dipolar shift tensor $\tilde{\boldsymbol{\sigma}}$ can be treated as any other shift or shielding tensors [24], and can be decomposed into rank 0, rank 1 and rank 2 irreducible tensors [25, 26]. The rank 0 component correspond to the spatially averaged pseudocontact shift (see eq. 2.22):

$$\tilde{\sigma}^{[0]} = -\frac{1}{3}\,\mathrm{Tr}\,(\tilde{\boldsymbol{\sigma}}) = \frac{1}{4\pi}\frac{(3x^2-r^2)\,\chi_x + (3y^2-r^2)\,\chi_y + (3z^2-r^2)\,\chi_z}{3r^5} = \delta^{pc}$$

(2.52)

while the rank 2 irreducible tensor, responsible for relaxation [25, 26], is given by:

$$\tilde{\sigma}^{[2]} = \frac{1}{4\pi r^5}$$
$$\times \begin{pmatrix} (3x^2-r^2)\chi_{xx} - 4\pi r^5\delta^{pc} & 3xy(\chi_{xx}+\chi_{yy})/2 & 3xz(\chi_{xx}+\chi_{zz})/2 \\ 3xy(\chi_{xx}+\chi_{yy})/2 & (3y^2-r^2)\chi_{yy} - 4\pi r^5\delta^{pc} & 3yz(\chi_{yy}+\chi_{zz})/2 \\ 3xz(\chi_{xx}+\chi_{zz})/2 & 3yz(\chi_{yy}+\chi_{zz})/2 & (3z^2-r^2)\chi_{zz} - 4\pi r^5\delta^{pc} \end{pmatrix}$$

(2.53)

Assuming an isotropic susceptibility tensor, we have:

$$\chi_{xx} = \chi_{yy} = \chi_{zz} = \chi_{ave} = \mu_0\mu_B^2 g_J^2 \frac{J(J+1)}{3kT}$$

(2.54)

where μ_B is Bohr's magneton, g_J the metal g factor, J the metal spin, k Boltzmann's constant and T the absolute temperature; g_J and J are used instead of g_e and S for lanthanides. The shift tensor becomes axially symmetric, with:

$$\Delta\sigma = \sigma_\| - \sigma_\perp = \frac{\mu_0}{4\pi}\mu_B^2 g_J^2 \frac{J(J+1)}{r^3 kT}$$

(2.55)

Calculating the energy levels, the transition frequencies and the transition probabilities for a system now described by the Hamiltonian 2.51, the expressions for the enhancement of the relaxation rates of spin I are [21, 27, 28]:

$$R_1^\chi = \frac{2}{5}\gamma_I^2 B_0^2 \Delta\sigma^2 \left(\frac{3\tau_r}{1+\omega_I^2\tau_r^2}\right)$$
$$= \frac{2}{5}\left(\frac{\mu_0}{4\pi}\right)^2 \frac{g_J^4\mu_B^4\gamma_I^2 J^2(J+1)^2 B_0^2}{r^6(3kT)^2}\left(\frac{3\tau_r}{1+\omega_I^2\tau_r^2}\right)$$
$$R_2^\chi = \frac{1}{5}\gamma_I^2 B_0^2 \Delta\sigma^2 \left(4\tau_r + \frac{3\tau_r}{1+\omega_I^2\tau_r^2}\right)$$
$$= \frac{1}{5}\left(\frac{\mu_0}{4\pi}\right)^2 \frac{g_J^4\mu_B^4\gamma_I^2 J^2(J+1)^2 B_0^2}{r^6(3kT)^2}\left(4\tau_r + \frac{3\tau_r}{1+\omega_I^2\tau_r^2}\right)$$

(2.56)

Corrections caused by an anisotropic susceptibility tensor have been shown to be usually negligible [28].

2.2.3   Relative relevance of the different contributions

It is generally expected [1, 29] and explicitly shown [30] that the contact contribution to the LIR (equation 2.49) is negligible compared to the dipolar terms in solution (equations 2.44 and 2.56). On the other hand, the Solomon and Curie mechanisms can both play an important role for the nuclear relaxation; their relative importance can be estimated comparing equations 2.44 and 2.49 on one side and equation 2.56 on the other (Table 2.3). An indicative ratio ρ between the two different contribution can be expressed as:

$$\rho\left(\frac{\chi}{S}\right) \simeq \frac{g_J^2 \mu_B^2 J(J+1) B_0^2}{(3kT)^2} \frac{\tau_r}{\tau_s} \tag{2.57}$$

so it is expected that Curie contributions will be larger the higher B_0, J and the τ_r/τ_s ratio. Furthermore, the occurrence of the Curie mechanism can be recognized through the temperature dependence of the linewidths, since the line broadening features an η/T^3 dependence, arising from the $1/T^2$ dependence of $\langle J_z \rangle$ indicated in equations 2.56 and from the η/T dependence of τ_r (equation 2.40).

The most useful information provided by Solomon and Curie mechanisms concerns experimental values that are related to the distance of the nucleus from the paramagnetic metal and to τ_c or τ_r. Actually:

- for large molecules such as proteins, the Curie term dominates R_2 and contributes roughly the same as the standard term to R_1 [1]. Distances can then be calculated from equation 2.56 for R_2^χ making use of estimates of τ_r from the Stokes-Einstein relationships; τ_s is not needed;

- for relatively small molecules such as those examined in chapter 3, the Curie and standard terms are comparable for both R_1 and R_2. The Curie contribution can be factored out due to the different coefficients for the spectral densities which enter expressions 2.56; in the fast motion limit:

$$R_{2,i} - R_{1,i} = \frac{const}{(3kT)^2} \cdot \frac{1}{r_i^6}\left(4\tau_r - \frac{3\tau_r}{1+\omega_I^2\tau_r^2}\right) \tag{2.58}$$

It is anyway difficult to obtain actual distances because the correlation times are not usually known with sufficient accuracy. However, both the dipolar and Curie contributions to the relaxation are proportional to r^{-6}, thus a *generalized* term:

$$R_1^{\text{overall}} = C\frac{1}{r^6}$$
$$R_2^{\text{overall}} = C'\frac{1}{r^6} \tag{2.59}$$

allows ratios of ion-proton distances to be calculated from the relaxation data for different protons without giving any estimate of τ_c.

			longitudinal relaxation								transverse relaxation							
			$B_0 = 7$ T ($\omega_H = 300$ MHz)				$B_0 = 14$ T ($\omega_H = 600$ MHz)				$B_0 = 7$ T ($\omega_H = 300$ MHz)				$B_0 = 14$ T ($\omega_H = 600$ MHz)			
			$\tau_r = 0.1$ ns		$\tau_r = 1$ ns		$\tau_r = 0.1$ ns		$\tau_r = 1$ ns		$\tau_r = 0.1$ ns		$\tau_r = 1$ ns		$\tau_r = 0.1$ ns		$\tau_r = 1$ ns	
Ln^{3+}	J	g_J	R_1^S	R_1^X	R_1^S	R_1^X	R_1^S	R_1^X	R_1^S	R_1^X	R_2^S	R_2^X	R_2^S	R_2^X	R_2^S	R_2^X	R_2^S	R_2^X
Ce	5/2	6/7	4.4	0.5	4.4	0.2	4.2	2.0	4.2	0.2	4.4	0.6	4.4	37.4	4.2	2.5	4.2	149
Pr	4	4/5	3.6	2.1	3.6	0.6	3.6	7.8	3.6	0.6	3.6	2.6	3.6	148	3.6	9.8	3.6	592
Nd	9/2	8/11	13.7	2.2	13.7	0.7	12.2	8.1	12.2	0.7	13.7	2.7	13.8	155	12.3	10.3	12.4	620
Sm	4	3/5	0.3	0.01	0.3	0.001	0.3	0.02	0.3	0.001	0.3	0.01	0.3	0.5	0.3	0.03	0.30	1.80
Gd	7/2	2	9500	52	3300	15.1	8600	188	830	15	11300	61.8	550000	3600	11000	237	550000	143000
Tb	6	3/2	116	117	116	34.0	99.0	610	100	34	117	139	117	8100	101	770	101	32500
Dy	15/2	4/3	134	168	134	48.8	116	610	116	49	134	200	135	11600	118	770	118	46500
Ho	8	5/4	117	166	117	48.2	101	600	105	48.2	118	197	118	11500	106	760	106	45800
Er	15/2	6/5	87.2	110	87.4	32.0	79	400	79.2	32.1	87.5	131	87.6	7600	79.9	505	80	30500
Tm	6	7/6	74.3	42.8	74.5	12.4	62.7	155	62.9	12.4	74.7	50.9	71.9	2960	64	196	64.1	11800
Yb	7/2	8/7	16.4	5.5	16.4	1.6	15.3	20.1	15.3	1.6	16.4	6.6	16.5	380	15.4	25.3	15.4	1500

Table 2.3: Estimated longitudinal and transverse relaxation rate enhancements (R_i in s^{-1}) in lanthanide complexes. The values refer to the separate contributions of the Solomon and Curie mechanisms to the relaxation of a proton located at 5 Å from the metal, in the case of a small and a large molecule (tumbling correlation times $\tau_r = 0.1$ and 10 ns, respectively), at fields of 7 and 14 T and 298 K. The spin relaxation times from LnDTPA (Table 2.2) were used as τ_s.

In the case of Yb^{3+} complexes, like those studied in the present thesis, at 298 and at a field of $7.1T$, the equation for the longitudinal relaxation rates of a proton in s^{-1} in the fast motion limit take the form:

$$R_1^{\text{overall}} = \frac{1}{r^6} \left(1.69 \cdot 10^6 \tau_s + 8.85 \cdot 10^2 \tau_r \right) \tag{2.60}$$

where τ_s and τ_r are expressed in ps and r in Å.

2.3   Choice of the lanthanide ion

The choice of the best lanthanide ion for a NMR analysis depends from case to case on the kind of problem:

- gadolinium has the most slowly-relaxing highest magnetic moment and its complexes are ideal candidates as contrast agents in MRI [31]. In the most used complexes, the ion is tightly chelated by a macrocycle, while a bound water molecule can exchange rapidly with the bulk water in the solution; this process provides an efficient mechanism for the enhancement of the relaxation of the water protons and consequently of the intensity of the MRI signals of the regions of the body where the complex can distribute;

- ytterbium is one of the best qualified lanthanide ions for the structural investigations of small molecules by ^{1}H-NMR [32–34]; in these cases, most of the relevant nuclei surrounding the ion are closer than 5 Å, and this ion is the only one for which the LIS effect may be, with good approximation, totally ascribed to the pseudocontact term in such conditions;

- dysprosium is the best choice for the structural analysis of large molecular complexes or for the detection of outer sphere interactions between stable metal complexes and other substrates: this ion produces indeed the largest pseudocontact shift among the lanthanides (Bleaney factor $C_j = 100$). Tulium [35–37] and terbium [38] are other favourable ions for this kind of studies: even if they feature lower absolute Bleaney factors, thanks to a lower effective magnetic moment, induce smaller susceptibility effects and cause less line-broadening;

- Ca^{2+} replacement with lanthanides in calcium-binding proteins [39] offers a way for the solution refinement of such macromolecules, if induced pseudocontact shifts and dipolar relaxation effect are used as constraints [40, 41]; the shifting ability of ytterbium and dysprosium is counterbalanced by the great LIR effects, which hinders the resolution and the detection of cross-peaks in 2D spectra. This is the reason for which cerium is often preferred in this studies, since its ratio between induced shifts and paramagnetic broadening is most favourable. In a recent paper

[42], Ce^{3+}, Yb^{3+} and Dy^{3+} have been used to *enlighten* shells at variable distances from the metal in monolanthanide-substituted calbindin, with cerium providing an effective refinement in the range 5–15 Å, ytterbium in the range 9–25 Å and dysprosium in the range 13–40 Å;

- the orientation of paramagnetic metalloproteins in a magnetic field is determined by the extent of the magnetic anisotropy of the ion which, for a given ligand geometry, depends on the magnetic moment value of the ion; in some recent studies, ytterbium [43, 44] and dysprosium [44] have been employed, but again cerium ions seem to be preferred for the most favourable ratio between orienting capability and induced line broadening [45].

Bibliography

[1] BERTINI, I.; LUCHINAT, C., *Solution NMR of paramagnetic molecules. Applications to metallobiomolecules and models;* ELSEVIER: AMSTERDAM, 2001.

[2] BLEANEY, B., *J. Magn. Reson.* **1972**, *8*, 91–100.

[3] VAN VLECK, J. H., *The theory of electric and magnetic susceptibilities;* OXFORD UNIVERSITY PRESS: LONDON, 1932.

[4] ASHCROFT, N. W.; MERMIN, N. D., *Solid state physics;* HOLT-SAUNDERS: PHILADELPHIA, 1981.

[5] MCGARVEY, B. R., *J. Chem. Phys.* **1976**, *65*, 955–961.

[6] MCGARVEY, B. R., *J. Chem. Phys.* **1976**, *65*, 962–968.

[7] MCGARVEY, B. R., *J. Magn. Reson.* **1979**, *33*, 445–455.

[8] BABUSHKINA, T. A.; ZOLIN, V. F.; KORENEVA, L. G., *J. Magn. Reson.* **1983**, *52*, 169–181.

[9] BERTINI, I.; YANIK, M. B. L.; LEE, Y.-M.; LUCHINAT, C.; ROSATO, A., *J. Am. Chem. Soc.* **2001**, *123*, 4181–4188.

[10] VIGOROUX, C.; BELORIZKY, E.; FRIES, P. H., *Eur. Phys. J.* **1999**, *5*, 243–255.

[11] MIRONOV, V. S.; GALYAMETDINOV, Y. G.; CEULEMANS, A.; GÖRLLER-WALRAND, C.; BINNEMANS, K., *Chem. Phys. Lett.* **2001**, *345*, 132–140.

[12] MIRONOV, V. S.; GALYAMETDINOV, Y. G.; CEULEMANS, A.; GÖRLLER-WALRAND, C.; BINNEMANS, K., *J. Chem. Phys.* **2002**, *116*, 4673–4685.

[13] SHERRY, A. D.; GERALDES, C. F. G. C., , IN *Lanthanide Probes in Life, Chemical and Earth Sciences* BÜNTZLI, J.-C.; CHOPPIN, G. R., EDS.; ELSEVIER: AMSTERDAM, 1989 .

[14] REUBEN, J.; ELGAVISH, G. A., *J. Magn. Reson.* **1980**, *39*, 421–430.

[15] REUBEN, J., *J. Magn. Reson.* **1982**, *50*, 233–236.

[16] PLATAS, C.; AVECILLA, F.; DE BLAS, A.; GERALDES, C. F. G. C.; RODRIGUEZ-BLAS, T.; ADAMS, H.; MAHIA, J., *Inorg. Chem.* **1999**, *38*, 3190–3199.

[17] RIGAULT, S.; PIGUET, C.; BÜNTZLI, J.-C. G., *J. Chem. Soc., Dalton Trans.* **2000**, 2045–2053.

[18] RIGAULT, S.; PIGUET, C., *J. Am. Chem. Soc.* **2000**, *122*, 9304–9305.

[19] PIGUET, C.; EDDER, C.; RIGAULT, S.; BERNARDINELLI, G.; BÜNTZLI, J.-C. G.; HOPFGARTNER, G., *J. Chem. Soc., Dalton Trans.* **2000**, 3999–4006.

[20] SOLOMON, I., *Phys. Rev.* **1955**, *99*, 559–565.

[21] KOENIG, S. H., *J. Magn. Reson.* **1982**, *47*, 441–453.

[22] ALSAADI, B. M.; ROSSOTTI, F. J. C.; WILLIAMS, R. J. P., *J. Chem. Soc., Dalton Trans.* **1980**, 2151–2154.

[23] BLOEMBERGEN, N., *J. Chem. Phys.* **1957**, *27*, 572–573.

[24] BERTINI, I.; KOWALEWSKI, J.; LUCHINAT, C.; PARIGI, G., *J. Magn. Reson.* **2001**, *152*, 103–108.

[25] ANET, F. A. L.; O'LEARY, D. J., *Concepts Magn. Reson.* **1991**, *3*, 193–214.

[26] ANET, F. A. L.; O'LEARY, D. J., *Concepts Magn. Reson.* **1991**, *4*, 35–52.

[27] GUERON, M., *J. Magn. Reson.* **1975**, *19*, 58–66.

[28] VEGA, A. J.; FIAT, D., *Mol. Phys.* **1976**, *31*, 347–355.

[29] BANCI, L.; BERTINI, I.; LUCHINAT, C., *Nuclear and Electron Relaxation;* VCH: WEINHEIM, 1991.

[30] KEMPLE, M. D.; RAY, B. D.; LIPKOWITZ, K. B.; PRENDERGAAST, F. G.; RAO, B. D. N., *J. Am. Chem. Soc.* **1988**, *110*, 8275–8287.

[31] CARAVAN, P.; ELLISON, J. J.; MCMURRY, T. J.; LAUFFER, R. B., *Chem. Rev.* **1999**, *99*, 2293–2352.

[32] AIME, S.; BOTTA, M.; ERMONDI, G., *Inorg. Chem.* **1992**, *31*, 4291–4299.

[33] DI BARI, L.; PINTACUDA, G.; SALVADORI, P., *Eur. J. Inorg. Chem.* **2000**, 75–82.

[34] DI BARI, L.; PINTACUDA, G.; SALVADORI, P.; PARKER, D.; DICKINS, R. S., *J. Am. Chem. Soc.* **2000**, *122*, 9257–9264.

[35] ZITHA-BOVENS, E.; VAN BEKKUM, H.; PETERS, J. A.; GERALDES, C. F. G. C., *Eur. J. Inorg. Chem.* **1999**, 287–293.

[36] SHERRY, A. D.; ZARZYCKI, R.; GERALDES, C. F. G. C., *Magn.*

Reson. Chem. **1994**, *32,* 361–365.

[37] CORSI, D. M.;   VAN BEKKUM, H.;   PETERS, J. A., *Inorg. Chem.* **2000,** *39,* 4802–4808.

[38] BIEKOFSKI, R. R.; MUSKETT, F. W.; SCHMIDT, J. M.; MARTIN, S. R.; BROWNE, J. P.; BAYLEY, P. M.; FEENEY, J., *FEBS Lett.* **1999,** *460,* 519–526.

[39] PIDCOCK, E.; MOORE, G. R., *J. Biol. Inorg. Chem.* **2001,** *6,* 479-489.

[40] LEE, L.; SYKES, B. D., *Biochemistry* **1983,** *22,* 4366–4373.

[41] BENTROP, D.; BERTINI, I.; CREMONINI, M. A.; FORSÉN, S.; LUCHINAT, C.; MALMENDAL, A., *Biochemistry* **1997,** *36,* 11605–11618.

[42] ALLEGROZZI, M.; BERTINI, I.; JANIK, M. B. L.; LEE, Y.-M.; LIU, G.; LUCHINAT, C., *J. Am. Chem. Soc.* **2000,** *122,* 4154–4161.

[43] MA, C.; OPELLA, S. J., *J. Magn. Reson.* **2000,** *146,* 381–384.

[44] VEGLIA, G.; OPELLA, S. J., *J. Am. Chem. Soc.* **2000,** *122,* 11733–11734.

[45] BERTINI, I.; JANIK, M. B. L.; LIU, G.; LUCHINAT, C.; ROSATO, A., *J. Magn. Reson.* **2001,** *148,* 23–30.

Part II

Inert lanthanide chelates

Chapter 3

C_4 symmetric complexes

3.1 Generalities on the DOTA-like systems

Pendant donor ligands derived from polyaza macrocycles can bind several metals in a thermodynamically stable and kinetically inert fashion, the ions being wrapped by the ring nitrogens and by further oxygen donors, belonging to side arm groups [1]. The cavity preformed by the ligand is endowed with low flexibility and exerts size selectivity in complexing metal ions. So, while molecules derived from 1,3,5-triazacyclononane are effective ligands for transition metals as Zn(II), V(IV), Cr(III), Cu(II), Ni(II), Mn(IV) [2], Co(II) and Mn(II) [3]; on increasing the ring size, macrocyclic octadentate ligands derived from a tetraazacyclododecane (cyclen) ring have received the attention of several research groups for their ability to form stable and non-labile complexes with alkaline [4–7], alkaline earth [8], lead [9], cadmium [9, 10] and, mainly, lanthanide [11] cations.

In particular, one of the most successful ligands, DOTA[1] (**1**, figure 3.1), whose Gd(III) complex is widely used as contrast agent in MRI and in biomedical applications, has been thoroughly investigated through the whole f-transition. [12–19]

In the last decade, a large number of DOTA derivatives have been developed, in which the carboxylate side arms have been extended, functionalized or replaced by carboxyamides, phosphinates or alcohols [11].

The solid-state X-ray characterization of Ln^{3+} complexes with DOTA-analogues reveals that the eight coordination sites, four nitrogens and four oxygens, provided by these ligands, are arranged in two parallel squares, tilted by an angle ϕ (figure 3.4)). According to the value of this angle the coordination polyhedron can range from a perfect antiprism ($\phi = 45°$) to a prism ($\phi = 0°$). Neither limiting case has been observed so far in the crystal structures and most known geometries cluster around two values: $\phi = 15°$, $30°$. For example belong to the latter LnDOTA (Ln=Eu [13], Gd [15], Lu [17]) and EuDOPEA [20]; to the former can be assigned LaDOTA, [19] LaDOTAM, [21] EuTHP, [22] Y

---

[1]Abbreviated names for this and the other compounds are given in appendix C.

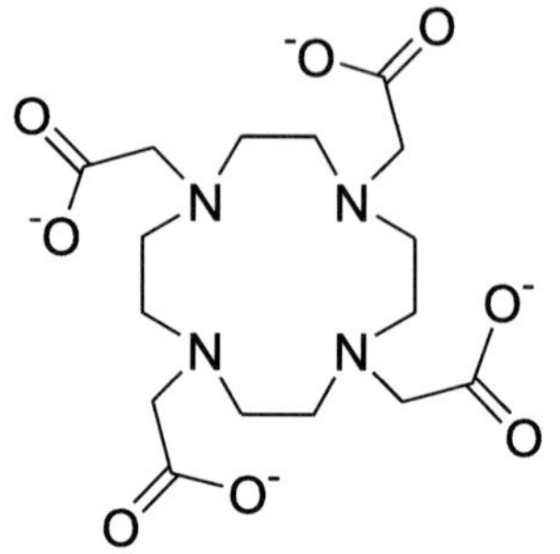

1

Figure 3.1: The *prototypical* DOTA ligand.

DOTPBz4 [23] and NaDONEA [24].

These two structures actually correspond to two main internal degrees of freedom. One concerning the conformation of the ring (figure 3.7a), which can assume two forms, referred to as $(\delta\delta\delta\delta)$ and $(\lambda\lambda\lambda\lambda)$ [25] and are in enantiomeric relation. The other involves the rotation about the N-C$^\alpha$ bond (figure 3.7b), switching the substituent residue between two *gauche* conformations. This induces a sign change in the distortion of the coordination polyhedron, named as Δ or Λ.

The two descriptors can be alike or different, e.g. $\Lambda(\lambda\lambda\lambda\lambda)$ or $\Lambda(\delta\delta\delta\delta)$, giving rise to a diastereomeric pair p or n, respectively (figure 3.3, R=H, Me).[2] A further relevant parameter to take into account is the orientation of the carboxylate bond, which is less accessible to experimental determinations via NMR.

The two diastereomeric forms n and p show a different arrangement of the oxygen atoms, in order to preserve the Ln-O distances at a value of about 2.4 Å. Accordingly, an almost cubic ($\phi = 22°$) coordination polyhedron, corresponds to the p diastereomer, while the n diastereomer has an almost antiprismatic ($\phi = 40°$) structure. [14, 18] To avoid confusion, table 3.1 provides the correspondence between the nomenclature rules employed in this thesis and other widespread names used in recent years to designate the diastereomeric pairs.

Most complexes have also been structurally characterized in solution, often using the paramagnetic character of the metal centre. [12–14, 16, 18, 27] While in most cases there is no evidence of structural equilibria and X-ray geometries possibly reflect the most likely conformation in solution, the complexes of DOTA reveal an intriguing network of solution equilibria, which have been studied by various NMR techniques, with reference to limiting crystal geome-

---

[2]We prefer to borrow the stereochimical descriptors n and p, which clarifies the diastereomeric nature of the rearranged forms, and avoids reference to the major form of YbDOTA. Indeed, the structures of the predominant form in solution depends on the nature of the ligand and on the size of the central ion.

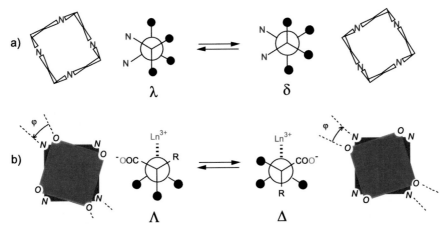

Figure 3.2: a)Equilibria between the δ and λ form of the cyclen ring (see ref. [21]). The Newmann projection is referred to the bond connecting the two ethilenic carbons of the macrocycle; the black circles represent the hydrogen atoms; b) equilibria between the Δ and Λ conformtions of the cyclen side arms in the DOTA-type complexes with Ln(III). The Newmann projection refers to the bond between the acetate C^α and the nitrogen; the black circles represent the hydrogen atoms, Y=O,N and R=H,Me. At the bottom, a schematic representation of the coordination polyhedron is depicted: at the corners of the light gray squares are the oxygens and at the corners of the dark ones the nitrogen atoms (the water molecule axially coordinated is not indicated).

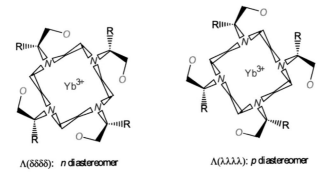

$\Lambda(\delta\delta\delta\delta)$: *n* diastereomer $\Lambda(\lambda\lambda\lambda\lambda)$: *p* diastereomer

Figure 3.3: Diastereoisomeric pair in DOTA-type complexes.

tries. These processes were thoroughly investigated by Aime and coworkers, who pointed out that the ratio of populations of *n*- and *p*-types varies along the lanthanide series. [14] In the case of DOTA complexes, superimposed to the conformational change, there is a hydration-dehydration equilibrium, [18] whose position depends on the size of the central ion. [14] In the case of achiral amide macrocycles [28, 29], this latter equilibrium has recently been shown to be dependent on the conformation of the macrocycle, with a square

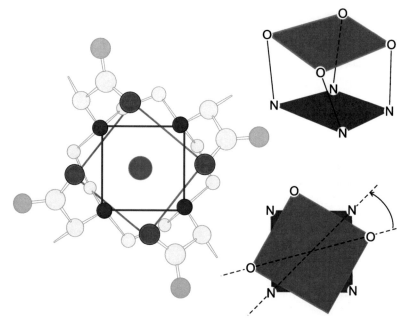

Figure 3.4: Schematic representations of the Ln(III) coordination polyhedra in DOTA-like systems.

	$\Lambda(\lambda\lambda\lambda\lambda)$ or $\Delta(\delta\delta\delta\delta)$	$\Lambda(\delta\delta\delta\delta)$ or $\Delta(\lambda\lambda\lambda\lambda)$	
tilt angle ϕ	22°	40°	
stereochemical descriptor	p-form	n-form	[26]
isomer in EuDOTA conformation equilibrium	m (minor form)	M (major form)	[18]
distortion of the coordination polyhedron	TSA (twisted square antiprism)	SAP (square antiprism)	[18]

Table 3.1: Different names adopted for the diastereomeric pairs in DOTA-like systems.

anti-prismatic coordination complex affording lower rates than a distorted (or twisted) anti-prismatic one.

3.2   Chiral analogues

Chiral Ln^{3+} complexes are expected to show optical activity of the lanthanide ff transitions owing to the dissymmetrical crystal field the cation experiences. Indeed, CD in the near infrared (980 nm) of Yb complexes has been preliminarily reported. [30–32] To understand the electronic spectroscopy of the lanthanide transitions, it is mandatory to have a clear picture of the structure in solution, particularly regarding the coordination polyhedron. Also, all the conformational rearrangements occurring at ordinary temperatures must be fully characterized. We undertook a thorough conformational and dynamical investigation of a set of chiral molecules, making profit of the huge amount of work on DOTA-like molecules. [12–19]

3.2.1 YbDOTMA

In the past fifteen years years, many chiral non-racemic derivatives and homologues of DOTA have been synthetized and studied. [20, 24, 27, 33, 34] Among them, DOTMA, **2**, (figure 3.5) is one of the most promising ligands for the high relaxivity of its Gd complex. [35] Its crystal structure has been solved and some data reported, but the full set of coordinates is not yet available. [35] The pseudocontact shifts of YbDOTA first reported by Desreux [12] reveal that the two sets of signals for the two forms are proportional. This means that the sets of geometrical factors of the protons in the two forms ($\frac{3\cos^2\theta-1}{r^3}$ in eq. 2.23) are equal. This does not necessarily mean that each proton occupies a position characterized by the same geometrical factor in the two forms, since there can be a pairwise proton-swap between equivalent sites. In 1992, Aime and coworkers [14] assigned the two spectra to two different structures, corresponding to the diastereomeric pair p and n. Indeed, they demonstrated the occurrence of such a rearrangement through EXSY measurements, as discussed in ref.s [14] and [16] and below. However, no positive evidence of the assignment of the n structure to the major component of YbDOTA in solution can be found in the pseudocontact shifts. On the contrary, such a simple experiment as a monodimensional NOE measurement can be employed to unequivocally assign the structure of the major isomer.

The presence of a stereogenic center on the side arm in DOTMA complicates the equilibrium network, since, for instance, the two forms $\Lambda(\lambda\lambda\lambda\lambda)$ and $\Delta(\delta\delta\delta\delta)$ are no longer enantiomers. The intrinsic chirality of the ligand is shown to induce a preference in the distorsion of the coordination polyhedron, which is demonstrated to be Λ. This finding is essential in understanding the chiroptical properties of the molecule.

3.2.2 YbDOPEA and YbDONEA

Chiral carboxamide ligands DOPEA (**3**) and DONEA (**4**) have been extensively investigated in the past two years and several reports have appeared on their complexes along the lanthanide series [20, 24, 36–39].

In the case of YbDOTMA, it has been shown that the presence of chiral substituents on the lateral chains inhibits the $\Lambda \rightleftharpoons \Delta$ interconversion by freezing

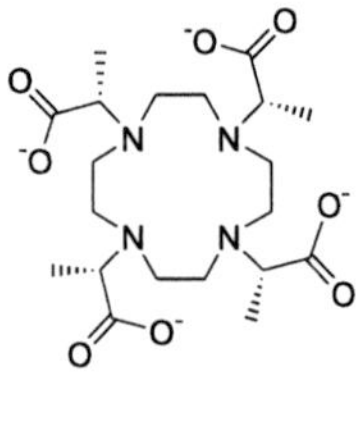

(R)-DOTMA
(R)-**2**

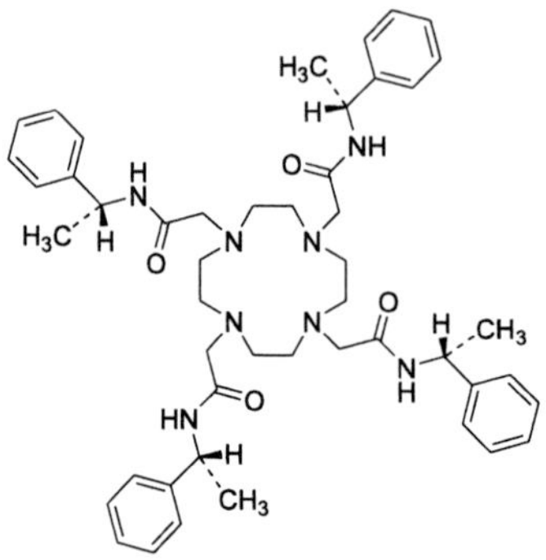

(R)-DOPEA
(R)-**3**

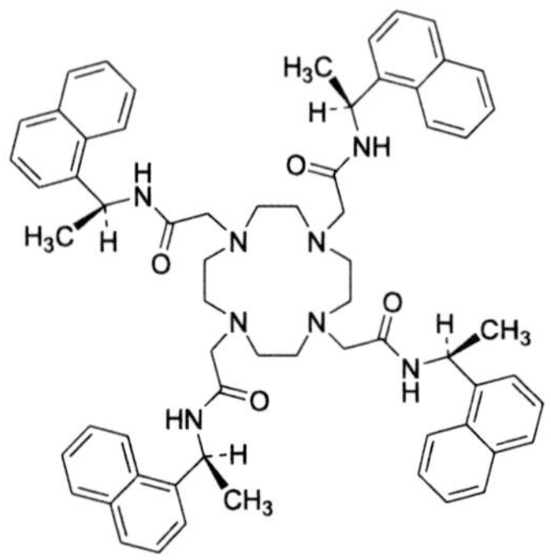

(S)-DONEA
(S)-**4**

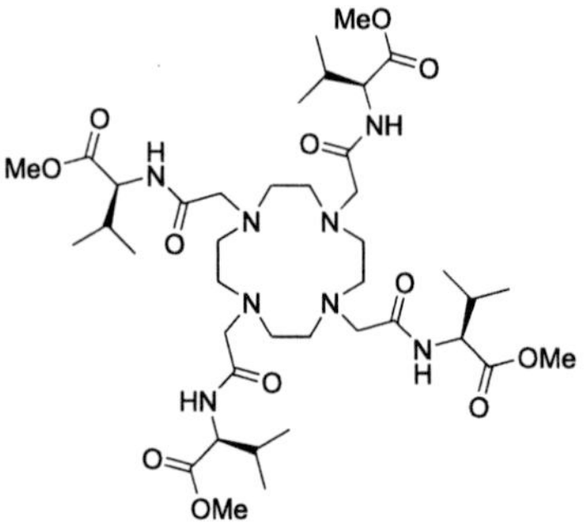

(S)-DOTAM-Ile
(S)-**5**

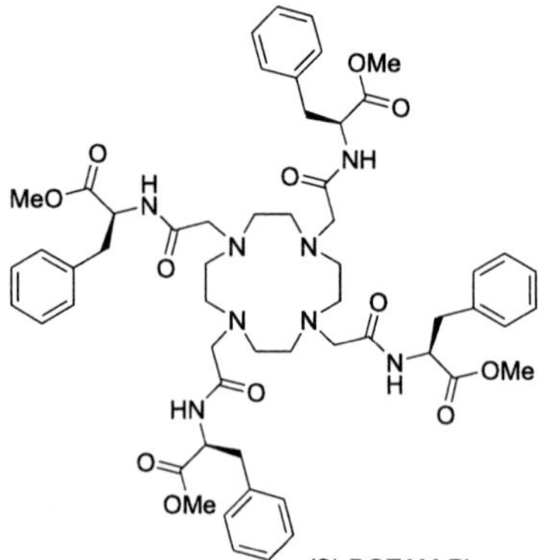

(S)-DOTAM-Phe
(S)-**6**

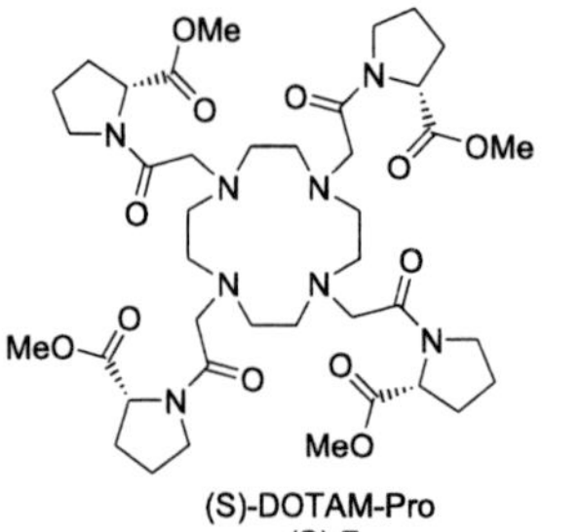

(S)-DOTAM-Pro
(S)-**7**

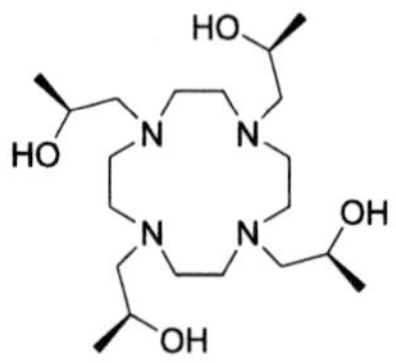

(S)-THP
(S)-**8**

Figure 3.5: Ligands studied in chapter 3.

the arm rotation [27, 33, 40]; if the centre is in δ position to the ring nitrogen, then only one isomer is commonly observed for complexes of Eu, Tb and Yb in solution [36]: thus the interpretation of the properties of the lanthanide complex is simplified by the absence of the conformational equilibria [41].

A preliminary study on the hydration properties and the axial coordination dynamics has recently been undertaken [37–39], but no full rationalisation has been put forward that accommodates all the available experimental data.

It will be discussed that for these complexes the coordination equilibrium involving a ninth axially bound ligand is independent of the ring dynamics, making it possible to assess the role of axial ligation on the main magnetic and electronic properties of the central cation (magnetic anisotropy susceptibility, crystal field parameters, absorption intensities). This is a matter of primary importance in understanding the solution behaviour of such complexes: the determination of the factors governing the effectiveness of ligand exchange may, in turn, offer new insight into the design of more effective contrast agents [42]. Furthermore, since the catalytic action of lanthanide complexes is often associated with a labile coordination equilibrium involving one or more reactants in the activated state of the reaction, information about the accessibility of the rare-earth ion may be of use in modulating the catalytic activity, in rationalising the preferred reaction pathways and in understanding the degree of stereoinduction [43].

As a starting point for the discussion of the solution coordination properties of these complexes, the helicity of the ring and of the side arms needs to be considered. Such an analysis begins with an examination of the solid state data: X-ray structures of Ln-**3**complexes have been reported in the last three years, for Ln=Eu, Dy, Yb [36, 39]. Although no crystal data are available for ligand **4** complexed to lanthanides, we can use as a reference the solid state geometry of the Na$^+$ complex of the same ligand (R)-**4** [24].

The X-ray structures reveal that Ln^{3+}-**3** and Na$^+$-**4** in the solid state have antiprismatic and distorted antiprismatic coordination (n- and p-forms) respectively, with the same layout of the acetate arms, leading to a Λ coordination, and the conformation of the macrocycles being $\delta\delta\delta\delta$ for Ln^{3+}-**3** and $\lambda\lambda\lambda\lambda$ for Na$^+$-**4**.

Structural assignment in solution by the analysis of the pseudocontact ^{1}H-NMR shifts needs special care, because:

- it has been reported in the case of YbDOTA and YbDOTMA that n- and p-forms can lead to practically equal geometrical factors [40];

- all the protons which exhibit distinct pseudocontact shifts lay within an inner sphere (i.e. in the DOTA-portion of the molecule). As a consequence, no direct relation can be found to the chirotopic elements of known configuration (that is the amine carbons) and one cannot discern between pairs $\Lambda(\delta\delta\delta\delta)$ and $\Delta(\lambda\lambda\lambda\lambda)$ for the antiprismatic coordination or $\Lambda(\lambda\lambda\lambda\lambda)$ and $\Delta(\delta\delta\delta\delta)$ for the distorted antiprism.

It is therefore of prime importance firstly in the case of Yb·(R)-**3**, to confirm the persistence in solution of the conformation observed in the crystal phase; secondly in the case of Yb·(S)-**4**, to verify whether the Na$^+$ solid state complex is a good description of the geometry in solution.

Once detailed information on the two coordination polyhedra is available, then a safe analysis of the magnetic and chiroptical properties of these species is made possible. While carrying out this purely structural investigation, involving the combined use of NMR and electronic spectroscopy, the key role of axial ligation on the magnetic and electronic properties of ytterbium became evident.

A variation in the NMR spectral features of lanthanide complexes following different axial bindings has been recently detected [44–46]. This was attributed to a change in the magnetic susceptibility anisotropy, following the modification (or the expansion) of the coordination sphere. Unfortunately, it is difficult to exclude in general a simultaneous readjustment of the ligand structure [44, 46]. In the only reported example involving a rather rigid macrocycle, changes in the effective crystal field parameters accompany the exchange of the anionic counterions [45]. We here take into account coordination with neutral solvent molecules. Such a process is likely to be amost ubiquitous in the solution chemistry of lanthanides, which are characterized by a large and variable coordination number. It follows a general rationalisation of a number of phenomena that may be observed by studying these complexes in different solvents.

3.2.3   YbDOTAM-Phe, YbDOTAM-Ile and YbDOTAM-Pro

Analogues of Yb·(R)-**3** and Yb·(S)-**4** can be easily synthetised employing natural amino acids as building blocks for the ligand side arms.

Molecules Yb·(S)-**5**, Yb·(S)-**6** and Yb·(S)-**7** were obtained as described in the experimental section from phenylalanine, isoleucine ans proline, respectively.

Even if no data exhist relative to the structure of these complexes either in solution or in the solid state, a complete analysis can be carried out exploiting the information coming from the analogues Yb·(R)-**3** and Yb·(S)-**4**.

The presence on the side arms of bulkier and more complex substituents will give us the chance of checking how the portion of the ligand lying outside the first coordination sphere affects the magnetic and electronic properties of the central cation.

3.2.4   YbTHP

The encapsulation of the ion and the formation of the complex at the same time affects the macrocycle characteristics and modulates the metal ion coordination properties.

In the case of aza-rings functionalized with hydroxyl-containing side arms, the coordination of the metal increases the acidity of the OH groups [47], while favouring a particular positioning of the side arms [9]. As a consequence, some of these molecules can behave as molecular receptors, forming H bridges with different substrates, and notably homo- (e.g.: Co(III) [48]) and hetero-dimers (e.g.: Mn(II)/Mn(IV) [3] or V(IV)/Zn(II) [49]) or supramolecular adducts [9], with promising applications in host-guest chemistry. On the other hand, the macrocyclic ligand rules the accessibility to the metal ion, which, according to its size and to the nature of the cage, can further coordinate (axially) another species [11]. These two reciprocal influences between central ion and ligand are at the origin of the use of such complexes in catalysis [43]. To quote only a few examples, a chiral triazacyclononane with alcoholic pendant groups complexed to Mn(II) is active in the enantioselective olefin epoxydation [50]; lanthanide complexes with ligands such as THED [47, 51, 52] or its chiral analogue THP **8** [22, 47] have proved effective in promoting transesterification of phosphate diesters and cleavage of RNA oligomers [47, 53]; furthermore, the same complexes are under investigation in the attempt of replacing lanthanide alkoxides in the Meerwein-Ponndorf-Verley reduction [54–56], with possible enantiose-lective conversions when chiral macrocycles are employed. In all kinds of reactions, the catalytic activity may be rationalized as based on two steps (figure 3.6) [47, 53, 57]: the deprotonation of a lanthanide-bound hydroxyl group or a lanthanide-bound water molecule, which can act as a general base catalyst, and the axial coordination of a substrate molecule (e.g.: phosphate).

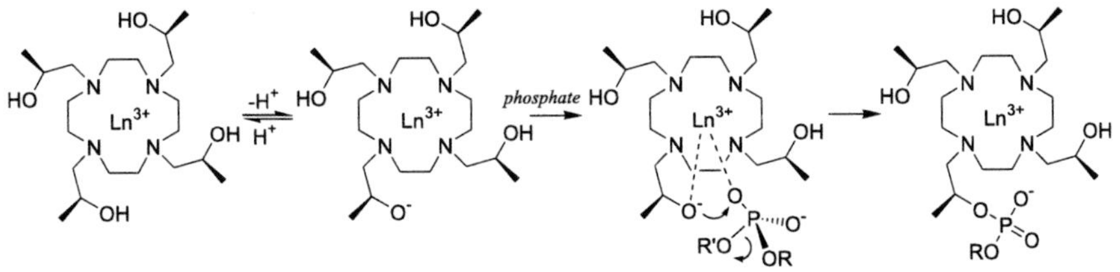

Figure 3.6: Hypothesis for the mechanism of the LnTHP promoted RNA transesterification (according to ref. [47, 53]).

Moreover, the "controlled" accessibility of the central lanthanide cation is the key point of all the numerous uses rare earths have been experiencing in the last years in different fields of chemistry, biology and medicine, as luminescent chelates [58–60] for immunoessays, as shift reagents in NMR [61] and as contrast agents in diagnostic medicine [11, 42, 62].

Major pieces of information about the structure of such metal complexes is provided by the X-ray analysis in the solid state [21, 22], due to the lack of techniques capable of providing a detailed description of the solution chemistry. The data available on the solution conformation of these molecules rely on NMR experiments (mainly on diamagnetic compounds of La and Lu [52] or their alcaline [4–7] and alcaline earth [8] analogues), which allow one to detect the number of different species in slow exchange and to characterize the thermodynamics and kinetics of such processes, possibly (for alcalines and alcaline earths) coupling the analysis to thorough *ab initio* calculations [6, 7, 9]. While on the analogue derivatives of the tri-membered aza-rings with Mn^{2+}, Ni^{2+} and Co^{3+}, absorbance and CD studies on the central cation have been performed [3, 48, 49], electronic spectroscopy has rarely been used in the case of lanthanide derivatives. Only luminescence studies on europium complexes

[57, 63], have been employed to check, in a qualitative fashion, the presence of different species in solution and to detect possible solvent and substrate bindings.

It is anyway noteworthy that, on such substrates,the lanthanide-induced NMR shifts and relaxation and the absorbances and CDs of the f-f transitions have never been exploited.

Homochiral THP (**8**), complexed to La, Eu and Lu, is present in solution as a single species [22], since the single chirality on the side arms amplificates in a single configuration of the coordination polyhedron and, at the same time, freezes the ring dynamics.

Published X-rays data of the Eu complex [22] do not offer a univocal picture of the system, since crystals were collected from a stereomeric mixture: together with the homochiral complexes (*RRRR* and *SSSS*), two other species cocrystallized in the unit cell, with *RRRS* and *SSSR* configurations at the chiral centres. While the former diastereomers adopt a *p*-type coordination ($\Lambda(\lambda\lambda\lambda\lambda)$ and $\Delta(\delta\delta\delta\delta)$), the latter exhibit a *n*-type form ($\Delta(\lambda\lambda\lambda\lambda)$ and $\Lambda(\delta\delta\delta\delta)$). In both cases, the cation is nine-coordinated, with an additional water molecule capping the O_4 face of the primary coordination sphere.

As far as solution is concerned, the only dynamics detected involves a protonation-deprotonation equilibrium, which can be assigned to a bound water molecule or to a coordinated hydroxyl moiety [47, 53]; the pK_a, measured by potentiometric titrations, decreases on traversing the transition and follows the order $La^{3+}>Eu^{3+}>Lu^{3+}$, in opposite trend with respect to the catalytic efficiency in the phosphate transesterification; the Lu complex has a second pK_a at around 9.3, which may be attributed to the greater Lewis acidity of such ion.

3.3   Description of the method

In the following part of this chapter, we shall face a stepwise structural analysis of this set of molecules: problems concerning different aspects of the solution geometry and coordination will be solved in different sections and we will discover which different experimental techniques can prove more suitable in each case.

As anticipated in the introduction, the problems concerning these macrocyclic complexes are:

the geometry of the coordination cage: ¹H-NMR is the most powerful technique for the determination of the portion of the molecule most paramagnetically shifted (< 5 Å from Yb^{3+}) and allows the description of the polyhedron (square or twisted antiprism) around the metal.

The proton shift in a lanthanide complex can be written as:

$$\delta = \delta_{dia} + \delta_c + \delta_{pc} \qquad (3.1)$$

where the three contributions are the diamagnetic, paramagnetic contact and paramagnetic pseudo-contact terms respectively. In the case of Yb(III):

- the contact contribution is usually negligible (see paragraph 3.7 for a discussion);

- the diamagnetic reference term can be provided by the shifts of the Lu^{3+} analogues or by those of the free ligands.

In the case of axially symmetric complexes, the pseudo-contact shifts can be expressed as:

$$\delta_{pc} = \mathcal{D}\frac{3cos^2\theta - 1}{r^3} + \Delta \qquad (3.2)$$

where θ and r are the polar coordinates of a nucleus respect to the Yb(III) ion, $\mathcal{D}$ is the magnetic anisotropy factor, in units of ppm·$Å^3$, and a constant Δ can be introduced to take into account the shift of the reference signal caused by partial coordination of the standard *tert*-buthyl alcohol.

As described in section A.6.1, the experimental pseudocontact values are then fitted against a set of geometrical factors derived from a crystal or a model structure. These geometrical values are allowed to change according to the main degree of freedom of the molecule, till the best agreement is reached with the experimental data.

In some cases, only the aid of NOE data will provide unequivocal assignment of the conformations of the coordination polyhedra.

the chirality of the coordination cage: NIR-CD is the only experimental tool able to experience the sense of twist of the antiprismatic cage around the metal.

As discussed in section 1.4.3, this effect originates from the unique *ff* transition of Yb^{3+} between the $^2F_{7/2}$ and $^2F_{5/2}$ states, which are split by a C_4 symmetry crystal field in 4 and 3 doubly degenerates sublevels, respectively.

Owing to the small energy separation between the states of the lower term (of the order of 100 cm^{-1}), not only the ground sublevel is populated and a temperature-dependent Boltzmann distribution among the four states of the $^2F_{7/2}$ has to be expected; thus, up to 12 transitions might be observable about the center of gravity. The latter must be around 980 nm, as observed for the free ion, as it can not be altered by the crystal field splittings.

The circular dichroism is readily recorded owing to the very favorable dissymmetry g-factor (up to about 0.25 at 946 nm). In fact, as discussed in section 1.2, the transition is globally magnetically allowed and it must just borrow some electric dipole moment (e.g. from a *4f-5d* transition) to give rise to non-vanishing rotational strength.

The signs and intensities of the sequence of bands apparent from the CD spectrum is related to the dissymmetric distribution of the atoms belonging to the organic ligand and in particular the four oxygen and four nitrogen donor atoms;

the chirality of the side arms: UV-CD is the technique which allows one to get a measure of the relative position and distortion of the side arms containing organic UV-absorbing chromophores;

the axial equilibria and dynamics: ^{1}H-NMR , NIR-CD, luminescence and ESI-MS measurements are found to provide convergent evidences about the presence and the extent of the solvent coordination involving the axial site on the top of the molecule.

3.4   NMR analysis: description of the coordination cage

3.4.1   YbDOTMA

Analysis of the pseudocontact shift

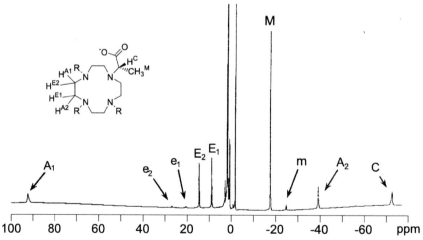

Figure 3.7: ^{1}H-NMR spectrum of YbDOTMA in D_2O at 25° C. The peak labelling is also indicated.

proton	$\left\langle \dfrac{3cos^2\theta - 1}{r^3} \right\rangle \times 100$	major form m-type, p diastereoisomer				minor form M-type, n diastereoisomer			
		D$_2$O		MeOD		D$_2$O		MeOD	
		δ^{exp}	δ^{calc}	δ^{exp}	δ^{calc}	δ^{exp}	δ^{calc}	δ^{exp}	δ^{calc}
axial$_1$	3.09	A$_1$ 94.3	94.0	109.9	109.8	a$_1$ 160.9	159.9		179.9
axial$_2$	-0.942	A$_2$ -38.9	-35.1	-41.1	-38.2	a$_2$ -55.9	-53.9	-61.0	-60.5
equatorial$_1$	0.502	E$_1$ 10.5	11.1	14.5	14.8	e$_1$ 22.2	22.7	25.8	25.6
equatorial$_2$	0.652	E$_2$ 16.3	15.9	20.5	20.3	e$_2$ 28.7	30.6	32.8	34.6
acetate^c	-2.16	C -72.5	-74.2	-81.8	-82.9	c -117.4	-118.5	-133.9	-133.1
methyl^c	-0.407	M -16.0	-18.0	-16.8	-18.6	m -23.2	-25.6	-25.6	-28.6
R^a, %		3.0		3.4		3.5		4.6	
$\mathcal{D}^b$		3200 ±90		3700 ±50		5300 ±100		6000 ±150	

^{a} Nonweighted agreement factor, see ref. [14] – ^{b} Magnetic anisotropy factor, in units of $[ppm \cdot \text{Å}^3]$ –
^{c} Average between the values found in the two forms obtained in the text (-2.09 and -2.23 for C;
-0.423 and -0.391 for M in the major and minor form, respectively).

Table 3.2: Experimental and calculated pseudocontact NMR shifts (ppm) of the major and minor YbDOTMA species in D$_2$O and MeOD (T=25°C); the protons are labelled according to figure 3.9. The factors in the second column are calculated on a structure derived from GdDOTA (see table 3.5 and ref. [15]); induced shifts are relative to diamagnetic LuDOTA (as far as it concerns the ring protons and the acetate one) and to a value of 0.9 ppm for the case of the methyl resonance.

The spectrum of YbDOTMA in water (Figure 3.7), as discussed by Brittain and Desreux, [27] shows that the complex is present in solution in two forms D and d in slow exchange, characterized by different anistropy factors $\mathcal{D}_D$ and $\mathcal{D}_d$ and different signal intensities, in analogy with YbDOTA. [14] However, unlike in DOTA, here the *less* populated form has the larger value of the $\mathcal{D}$ factor.

We can observe that the spectra of the two forms are exactly proportional (isomorphic), as demonstrated in figure 3.8, and only the large difference between the anisotropy factors, $\mathcal{D}_D$ and $\mathcal{D}_d$ justifies the appearence of two distinct sets of signals for the two isomers.

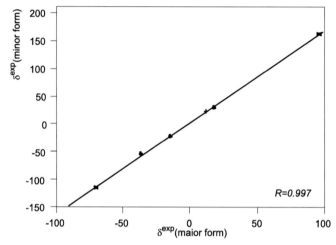

Figure 3.8: Correlation plot between the experimental δ^{exp} values of the major (D) and the minor (d) forms of YbDOTMA in D_2O.

It follows that, at least in principle, the two spectra can be associated to the *same* molecular structure. Alternatively, two conformations must be found bearing the protons in positions characterized by very similar geometrical factors.

Following Brittain and Desreux, [27] a set of suitable parameters for the spectral assignment can be produced from the coordinates of GdDOTA [15] (we chose to start from the $\Lambda(\delta\delta\delta\delta)$ form). The pro-R hydrogen on each acetate arm is replaced by a methyl group with standard bond lenghts and angles. The methylacetate arm is then rotated systematically around the N-C^α bond to obtain the best agreement[3] between the geometrical factors and the observed pseudocontact shifts, finding an optimal value of the torsion angle YbNC$^\alpha$Me of about 167° (which corresponds to a coordination geometry of Λ-type). The

---

[3]We can observe that, even if the R value for the major isomer is sligthly lower, the agreement is very good for both structures. Variation of the torsion around the N-C^α bond by 5° about the optimal value does not imply a relevant increase of the factor, R (about 10% higher when the angle is 170°).

carboxylate must then be rotated to restore the distance Yb-O of 2.4 Å, but this has no effect on the proton geometrical factors.

In such a way, a full set of satisfactory geometrical parameters and the relative structure are obtained. It should be observed that the solution of this optimization is unique, which excludes a rotation of the methylacetate as the origin of the two forms, as suggested by Aime. [34] This is in agreement also with other findings on similar structures. [33]

An equivalent set is provided by the following stereochemical operations on this structures:

1. mirror image (inversion) of the whole complex;

2. epimerization of the four C^α-stereocenters (these operations must be followed again by the carboxylate rotation).

The result is once more a (RRRR)-YbDOTMA but the arrangement of the macrocyclic ring is opposite. In other words, the tetraazacyclododecane shifts from a $(\lambda\lambda\lambda\lambda)$ to a $(\delta\delta\delta\delta)$ conformation. Obviously, starting from the $\Delta(\lambda\lambda\lambda\lambda)$ form of GdDOTA, one would obtain first the $\Delta(\lambda\lambda\lambda\lambda)$ isomer of YbDOTMA and, after the formal inversion and epimerization, the other, $\Lambda(\lambda\lambda\lambda\lambda)$. The two sets of geometrical factors for the cyclen ring are obviously equal. For the methyl group and for the CH such equality does not follow by necessity, however the difference between the geometrical factors calculated for the two forms is less than 10%. Any attempt to use this small difference for structural assignment is nullified by the proportionality of the spectra of the two species. Furthermore, smaller geometrical adjustments [14] can account for minor improvements of the agreement factors.

The geometrical parameters and the two relative structures are displayed in table 3.2 and figure 3.9, respectively.[4] The coordination geometry results of Λ-type in both cases, with the same torsion angles around the N-C^α bond; only the arrangement of the macrocycle is opposite.

Characterization of the dynamics

In order to determine the thermodynamic parameters for the isomerization equilibrium, a variable temperature study has been undertaken. On increasing the temperature the signals of the minor form become comparatively smaller, which indicates a negative enthalpy for this process. The spectra were recorded in methanol as well, in order to cover a broader temperature range. Indeed, the spectrum in this solvent is very similar to that in water reported in table 3.2; once more two forms are present, whose resonances can be predicted with the same geometrical factors used above. The two anisotropy factors are of the

---

[4]The geometrical factors calculated on the basis of the published of GdDOTA (ref. [15]) lead to the assignment shown in ref. [27], featuring the two protons on each macrocycle carbon bearing different numbers. This is in agreement with that reported by several authors (ref. [16, 25]), but in contrast with Aime and coworkers (ref. [14, 35]).

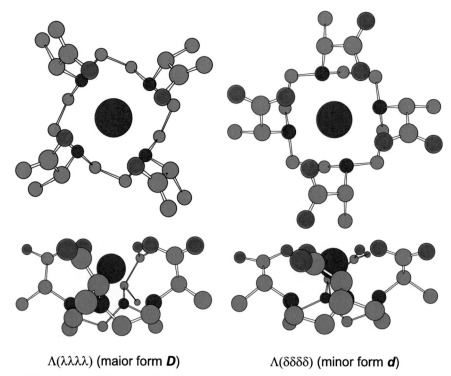

$\Lambda(\lambda\lambda\lambda\lambda)$ (maior form **D**) $\Lambda(\delta\delta\delta\delta)$ (minor form **d**)

Figure 3.9: Top and side views of the structure of the two isomers of Yb-DOTMA. On the left is the structure derived from GdDOTA, which corresponds to a $\Lambda(\lambda\lambda\lambda\lambda)$ conformation and has been assigned to the major isomer; the structure on the right ($\Lambda(\delta\delta\delta\delta)$, assigned to the minor isomer) has been obtained from the former by inversion of the macrocyclic ring.

same order of magnitude as those found in water and their ratio seems solvent-independent. Furthermore, in methanol the minor form is slightly more populated at room temperature than in water (D/d=12.4 and 20, respectively). The main difference between the two solvents is that in MeOD are clearly visible downfield shifted the resonances of the counter-ion, n-methylglucammonium, which are strongly concentration- and temperature-dependent. Therefore, we can conclude that the cation is at least partly bound in methanol, while it is not in D_2O. The downfield shift shows that it lies axially above or below the metal center.

On recording variable temperature spectra in MeOD between -80° and 60°C one observes a migration of all the signals according to the variation of $\mathcal{D}$, predicted by Bleaney. [64] Unfortunately, the values collected in table 3.3 do not allow to unambigously distinguish between a $1/T$ and a $1/T^2$ dependence. Furthermore, at low temperature, all the lines become broader, owing to faster transverse relaxation.

temperature (°C)	$\mathcal{D}$ (± 100)	$\frac{[d]}{[D]}$ %
60	3180	6.6
25	3490	7.0
0	3770	8.1
-25	4025	9.6
-50	4450	12.0
-80	5070	14.2

Table 3.3: Tmperature variation of the anisotropy susceptibility parameter $\mathcal{D}$ of the major form of YbDOTMA and of the equilibrium constant K between the two isomers.

The persistence of the signals of the minor form up to 60° could be observed, although at this temperature they appear very broadened. In fact, they start to coalesce with those of the major form, because of the faster equilibrium and of the smaller spectral width, which imply more stringent requirements for the slow exchange limit.

Integration of the signals of the methyl in the two forms affords an estimation of the temperature-dependence of the equilibrim constant, K, shown in table 3.3. This gives the thermodynamic parameters in methanol

$$\Delta H = -4.0 \pm 0.9 kJmol^{-1}$$
$$\Delta S = -35.4 \pm 0.5 JK^{-1}mol^{-1}.$$

In D_2O, the temperature range that can be covered goes from 0° to 50°, because above this limit the signals of the minor form become exceedingly broad. In water, the following values are found:

$$\Delta H = -8 \pm 2 kJmol^{-1}$$
$$\Delta S = -50 \pm 10 JK^{-1}mol^{-1}.$$

The dynamics of the interconversion between the two forms has been investigated through two-dimensional exchange spectra (EXSY) at variable mixing time in water and in methanol.

From inspection of figure 3.10, we observe the following equilibria, with reference to the notation of table 3.2:

$$M \rightleftharpoons m$$
$$E \rightleftharpoons a$$
$$A \rightleftharpoons e$$

while there is no evidence of exchange within the major or the minor form (e.g., $E \rightleftharpoons A$) , neither between equatorial (or axial) protons of the two forms (e.g., $E \rightleftharpoons e$). Therefore, we can conclude that the isomerization process involves a

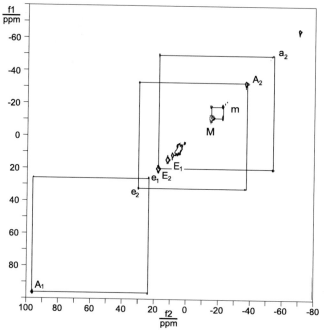

Figure 3.10: EXSY spectrum of YbDOTMA at 25°C in D_2O. The assignments of the peaks are taken from figure 3.8a.

ring inversion, exchanging the macrocycle hydrogens between positions characterized by pairwise equal geometrical factors.

Such an observation is in perfect agreement with the structure pair determined above and depicted in figure 3.9. Unfortunately, following their construction, it is impossible to assign either of the two to the major or minor component. To this aim, steady-state NOE [65] spectra were recorded between selected protons of the molecule.

When the resonance at -70.4 ppm, due to the methylacetate methine proton C is irradiated, an intense enhancement of E_2 is observed, while no effect is detected on E_1; at the same time, on saturating A_2, C is enhanced. The interpretation of these data is discussed in the next section.

A quantitative evaluation of EXSY spectra at variable mixing time, τ_m, allows one to assess the dynamic constant for the observed kinetic process. Direct analysis, as described by Perrin and Gipe, [66] of the integrated intensity matrix, $\mathbf{A}(\tau_m)$ for a given τ_m gives the dynamic matrix, $\mathbf{L}$:

$$\mathbf{L} = \frac{1}{\tau_m} \ln\left(\mathbf{P}^{-1}\mathbf{A}\right) = \frac{1}{\tau_m}\mathbf{U}^{-1}\ln(\Lambda)\mathbf{U}, \tag{3.3}$$

where $\mathbf{U}$ is the matrix of eigenvectors and the diagonal matrix Λ contains the eigenvalues, Λ_{ii}, of $\mathbf{P}^{-1}\mathbf{A}$. The diagonal matrix $\mathbf{P}$ contains the relative populations of the exchanging species, [67] so that the elements of $\mathbf{P}^{-1}\mathbf{A}$ are the normalized amplitudes of the peaks.

We repeated the experiment with different mixing times, ranging from 5 to 30 ms, and we measured the integral of the methyl signals in the two forms and of the cross-peaks connecting them, as reported in table 3.4.

		$\tau_m=5$ms	$\tau_m=10$ms	$\tau_m=20$ms	$\tau_m=30$ms
D_2O	$\begin{pmatrix} I_{11} & I_{12} \\ I_{21} & I_{22} \end{pmatrix}$	$\begin{pmatrix} 100 & 2.26 \\ 2.27 & 2.78 \end{pmatrix}$	$\begin{pmatrix} 100 & 4.40 \\ 4.70 & 1.72 \end{pmatrix}$	$\begin{pmatrix} 100 & 4.19 \\ 4.73 & 0.69 \end{pmatrix}$	$\begin{pmatrix} 100 & 4.72 \\ 5.32 & 0.33 \end{pmatrix}$
	$k_{d\rightarrow D}(M^{-1}s^{-1})$	6.0	7.5	5.3	6.4
	$k_{D\rightarrow d}(M^{-1}s^{-1})$	120.9	159.3	120.1	143.9

			$\tau_m=10$ms	$\tau_m=20$ms	$\tau_m=30$ms
MeOD	$\begin{pmatrix} I_{11} & I_{12} \\ I_{21} & I_{22} \end{pmatrix}$		$\begin{pmatrix} 100 & 5.1 \\ 6.4 & 5.4 \end{pmatrix}$	$\begin{pmatrix} 100 & 6.7 \\ 6.6 & 1.5 \end{pmatrix}$	$\begin{pmatrix} 100 & 7.8 \\ 8.4 & 1.4 \end{pmatrix}$
	$k_{d\rightarrow D}(M^{-1}s^{-1})$		6.3	7.7	6.6
	$k_{D\rightarrow d}(M^{-1}s^{-1})$		98.2	93.6	88.1

Table 3.4: Integrated intensity matrix for the exchange of the signals of the methyl group in water and in methanol. The index 1 represents the resonance frequency of M and 2 the one of m, so that I_{11} and I_{22} are the intensities of the diagonal peaks of the forms D and d, while I_{12} and I_{21} are the intensities of the cross-peaks, all normalized to $I_{11}=100$.

Cross-relaxation gives only a negligible contribution to the cross-peak amplitudes at the very short τ_m used in our experiments, hence the off-diagonal elements of $\mathbf{L}$ are $L_{ij} = -k_{ij}$, where k_{ij} is the first-order rate constant for the exchange from site i to site j.

Equation 3.3 then gives the rates also shown in table 3.4, whose average value is 6.3 ± 0.8 s$^{-1}M^{-1}$ for the D $\rightarrow$d process and 136 ± 16 s$^{-1}M^{-1}$ for the reverse in water, whereas, for the same processes in methanol one finds 6.9 ± 0.6 s$^{-1}M^{-1}$ and 93 ± 4 s$^{-1}M^{-1}$, respectively.[5]

These values seem to be quite reliable, since the ratio between the two constants in water (= 21.6) is in agreement with the relative concentrations of the two isomers (major/minor = 20 under our experimental conditions) and the same holds true in methanol, as well.

We can observe that the dynamics in the two solvents is very similar, which justifies us in extrapolating the thermodynamic data obtained in methanol to water.

---

[5]The errors on these values were exstimated through their dispersion for the various experiments.

Assignment of the structure

The possible solution equilibria in YbDOTMA can be represented, in analogy with YbDOTA, [18] as in figure 3.11.

The exchange processes in the rows are due to an inversion of the macrocyclic ring, while those on diagonals rely upon a rotation of the methylacetate arms. In the columns, the two motions occur simultaneously (and possibly concertedly). On the left the p diastereomers are presented, on the right the n. In the case of DOTA, the two forms along each column are enantiomers, whereas for DOTMA, the presence of the stereogenic centres at the C^α makes them diastereomers. Although this is only a crude representation, it shows that in the two rows the coordination polyhedra are distorted antiprisms with opposite tilt angles.

Clearly, since DOTA is achiral, the sign of such a distortion is irrelevant (it cannot be defined on a macroscopic level).

The populations of the four DOTMA structures must be expected different. Indeed, only two species could be identified in the monodimensional ^{1}H-NMR spectrum, in largely different amounts; both have Λ conformation of the side arm, bearing the methyl group *anti* with respect to N-Yb(III). The excellent agreement of the shifts with geometrical factors derived from a crystal structure rules out any fast motion and confirms the substantial rigidity of the chelate systems already reported in the literature. [14] There is no evidence of a rotation of the methyl acetate arm, which amounts to saying that only one of two rows of figure 3.11 is populated and therefore that only one equilibrium is present in solution for YbDOTMA, between the $\Lambda(\lambda\lambda\lambda\lambda)$ and $\Lambda(\delta\delta\delta\delta)$ forms. On the basis of the paramagnetic shifts only (as well as of Yb-proton distances and therefore of Curie or dipolar relaxation rates) it is not possible to assign either of the two conformations to the minor or major isomer, on the contrary, the NOE spectra allow one to assign univocally the structure of the major isomer to one of the two conformations. Indeed, the neat NOE between the C and E2 protons cannot be justified for an n-type structure, since the two protons are far apart (> 3.35Å). On the contrary, in the p-type the interproton distance is only 2.55 Å.[6] We can conclude that the $\Lambda(\delta\delta\delta\delta)$ must be assigned to the minor isomer and the $\Lambda(\lambda\lambda\lambda\lambda)$ to the major one, as depicted in figure 3.9; the relevant parameters of such geometries are shown in table 3.5. It must be stressed that, in reverse analogy with YbDOTA, the metal ion is more buried in the major form in the case of YbDOTMA.

The kinetic rates are of the same order of magnitude of those reported for Yb DOTA by Jacques and Desreux. [16] In the present case, however, only one process is observed, linked to the inversion of the macrocyclic ring. Apparently, the rotation of the methylacetate arm is strongly hindered, in analogy with what observed in some LnDOTA analogues. [33, 68] The detailed comparison

---

[6]The NOE between A2 and C is not diagnostic, since in both structures $\Lambda(\lambda\lambda\lambda\lambda)$ and $\Lambda(\delta\delta\delta\delta)$ the two protons are nearby. Furthermore, it is very difficult to detect, owing to the large relaxation rates of the two nuclei, which sit close to the paramagnetic center.

	major form	minor form
conformation	$\Lambda(\lambda\lambda\lambda\lambda)$	$\Lambda(\delta\delta\delta\delta)$
O – O	3.07	3.19
N – N	2.97	2. 97
Yb – O	2.48	2.28
Yb – plane (O_4)	1.01	0.72
Yb – N	2.66	2.66
Yb – plane (N_4)	1.63	1.63
twist angle (ϕ)	17 °	40°
YbNC$^\alpha$Me	167°	167°
NC$^\alpha$CO	324°	324°

Table 3.5: Conformation and relevant atomic distances [Å] and angles in the two isomers of ybDOTMA. The geometries were obtained from the X-ray data of GdDOTA [15], according to the procedure described in the text.

with the kinetic rates of DOTA is not straightforward, owing to the different molar ratios of the forms in the two complexes. The averages of the direct to reverse rates of ring inversion at 25°, weighted for the two populations gives 44.8 s^{-1} for Yb DOTA and 12.5 s^{-1} for Yb DOTMA. The dynamics of the latter is probably slowed down by the crowding due to the presence of the methyl on the side branches.

The thermodynamic data for the interconversion process show both negative enthalpy and entropy. Apparently, then, the minor form is slightly more stable and only entropically unfavored. The loss of entropy in the process D→d indicates that in this conversion the order of the system increases, which is compatible with a different coordination number of the central cation, that should be larger for the minor form. Aime et al. [18] reported that on going from the major to the minor form of Yb DOTA, ΔH^0=+17.5 KJ mol^{-1} and ΔS^0=+45 J K^{-1} mol^{-1} and they attributed the large and positive entropy change to the release of a water molecule, axially coordinated in the former. This hypothesis is strongly supported also by variable pressure-NMR. Since the reaction entropy is rather similar to what found for the conversion m' →M in DOTA, [18] it seems quite reasonable, then, to extrapolate this result to DOTMA, where it should be the major form the eight-coordinated one, while the minor should be hydrated and nine-coordinated; this further consideration is in agreement with the assignment of an M-type structure (n diastereomer) to the minor isomer, where the metal ion is less hindered.

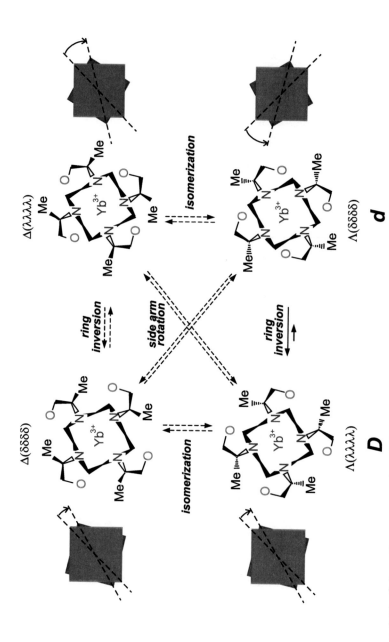

Figure 3.11: Schematic representation of the structure and dynamics of the stereoisomers of the YbDOTA (R=H) and YbDOTMA (R=Me) complexes. The conformation of the chelate systems and the coordination polyhedron are depicted. Following the interconcersion between the major and minor forms, a different distortion of the coordination polyhedron is forced in the two forms, as discussed in the text.

3.4.2 YbDOPEA and YbDONEA

Analysis of the pseudocontact shifts

The ^{1}H-NMR spectra of Yb·(R)-**3** and Yb·(S)-**4** in CD_3OD solution were measured and the assignment of the resonances was made from the relative peak areas and by comparison with the spectral data previously reported for Yb-DOTA [12]. Only the protons lying close to the Yb ion (distance from Yb^{3+} $r < 5$ Å) are significantly paramagnetically-shifted; they practically correspond to the DOTA-portion of the molecule, which is not involved in any observable dynamics over a wide temperature range (-40 to+40° C). The values of the chemical shift for each assigned signal are summarised in table 3.6 for Yb·(R)-**3** and Yb·(S)-**4**.

A set of reference geometrical factors for the analysis of the pseudocontact shifts is usually calculated from crystal structures. The X-ray structure of Yb·(R)-**3** (figure 3.12) provides the geometrical parameters corresponding to a square antiprism (*p*-type structure) [39]. At the moment, no crystal data are available for Yb(III) complexed in a distorted square antiprism (*p*-type geometry) as found in solution for YbDOTA (minor form) or YbDOTMA (major form). Two equivalent sets of polar coordinates θ and r, with reference to eq. 3.2, can be obtained:

- from the solid state geometry of Na·(R)-**4** [24] (figure 3.12);

- by manipulating the atomic coordinates of GdDOTA, with the aid of commercially available software for molecular graphics, by suitable stereochemical operations (e.g. rotation around the N-C^α bond, epimerization of the stereogenic centre at CC^δ, mirror imaging), as previously described for YbDOTMA [40].

The two sets of geometrical parameters obtained in this way and used for the optimisation are also reported in Table 3.6. It should be recalled that because the protons on the pendant arms exhibit small pseudocontact shifts, they were not used in structure calculation. The main degree of freedom for the DOTA-portion of the molecule is the torsion angle ω around N and C^α (defined here as the dihedral Yb-N-C^α-C^β shown in Figure 3.13). Thus a simultaneous fitting of ω and $\mathcal{D}$ was performed, starting from the two geometries for the *p*- and *n*-forms, as summarised in Table 3.6.

From inspection of table 3.6, it is apparent that no definite structural assignment for these molecules can follow from analysis of the pseudo-contact contribution only.

protons	$\left\langle \dfrac{3\cos^2\theta - 1}{r^3} \right\rangle \times 100$ n form (Eu·(R)-3)	p form (Na·(S)-4)	phenyl amide Yb·(R)-3 d4-methanol δ^{pc} exper.	δ^{calc} (n form)	δ^{calc} (p form)	d6-DMSO δ^{pc} exper.	δ^{calc} (n form)	δ^{calc} (p form)	naphthyl amide Yb·(S)-4 d4-methanol δ^{pc} exper.	δ^{calc} (n form)	δ^{calc} (p form)	d6-DMSO δ^{pc} exper.	δ^{calc} (n form)	δ^{calc} (p form)
a1	2.86	2.80	98.5	96.4	95.1	62.5	60.6	59.8	98.5	96.6	95.3	63.0	61.5	60.6
a2	-0.87	-0.92	-32.2	-29.6	-29.6	-19.3	-17.4	-17.2	-32.2	-30.1	-29.8	-18.8	-17.5	-17.3
e1	0.50	0.49	15.4	16.6	17.7	10.1	11.2	11.9	15.4	16.4	17.5	10.6	11.5	12.2
e2	0.63	0.61	18.4	21.0	21.8	12.0	13.9	14.5	18.4	20.8	21.7	12.5	14.2	14.8
c1	-0.85	-0.91	-27.0	-29.2	-29.1	-16.2	-17.1	-17.0	-27.9	-29.7	-29.6	-16.6	-17.3	-17.2
c2	-1.95	-2.06	-64.3	-66.3	-67.3	-37.9	-40.1	-40.7	-65.2	-67.0	-68.0	-38.3	-40.5	-41.1
m	-0.18	-0.15	-5.1	-6.2	-4.4	?	-3.3	-2.0	-5.4	-5.0	-4.6	-2.9	-3.3	-2.2
CH	-0.21	0.01	-3.8	-7.2	0.9	?	-4.0	1.3	?	-8.2	0.8	?	-4.0	1.1
NH	-0.72	-0.70	?	-24.6	-22.8	?	-14.9	-13.5	?	-25.6	-23.0	?	-14.8	-13.7
($\delta^{calc} - \delta^{pc}$) R (%) $\mathcal{D}$			2.6 0.26 3390 ± 70		3.6 0.47 3380 ± 95	2.1 0.45 2100 ± 60		2.7 0.72 2100 ± 70	2.3 0.19 3410 ± 60		3.2 0.38 3400 ± 90	2.3 0.65 2125 ± 45		1.7 0.35 2120 ± 60

Table 3.6: Experimental lanthanide induced shifts (δ, in ppm) of the Yb·(R)-**3** and Yb·(S)-**4** amide complex in d4-methanol, d3-acetonitrile and d6-DMSO solutions (T=25°C) and differences between the experimental pseudocontact contributions and those calculated in the fitting algorithm. The proton labelling is analogous to that commonly used for similar complexes (see table 3.3). The factors on the second and third columns are calculated from the crystal structures of Yb·(R)-**3** as described in the text and refer to structures which are representative of a n- type and a p-type, respectively. The experimental pseudocontact shifts used in the computations are the values relative to diamagnetic free ligands. Shifts in water are not reported, but they are analogue to those reported in the table for methanol. For each set of chemical shift fitted, the dihedral angle ω of the side chain, the agreement factor R and the susceptibility factor $\mathcal{D}$ are displayed.

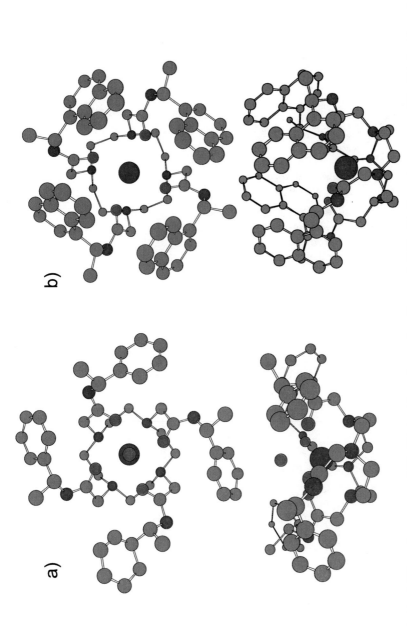

Figure 3.12: (a)Crystal structure of Yb·(R)-**3**, side and top view (from ref. [39]); (b) Crystal structure of Na·(R)-**4**, side and top view (from ref. [24]).

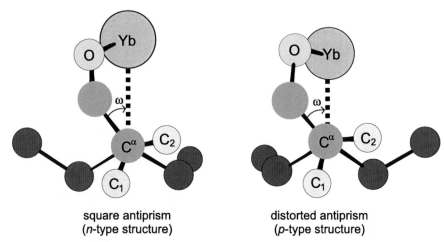

square antiprism   **distorted antiprism**
(*n*-type structure)   **(*p*-type structure)**

Figure 3.13: Position of the ring and of the side chains in the two different coordination types used as models for the fitting procedures. The pictures refer to a Newmann projection onto the N-C^α bond; the dihedral angle ω Yb-N-C^α-C is also displayed. In black are depicted the carbon atoms of the cyclen macrocycle.

The fact that both the *n*- and *p*-type geometries lead to very similar geometrical factors shows that no clear-cut information can be obtained from 1D ^{1}H-NMR spectrum as far as the conformation of the macrocycle is concerned. Such a conclusion has been noted in analysing the two forms of YbDOTMA [40]. Moreover, whilst the spectra are almost the same for the two complexes, the proportionality of the two sets of geometrical factors does not exclude the possibility that Yb·(R)-**3** and Yb·(S)-**4** adopt two diastereomeric structures. Indeed, the agreement factors are almost identical for both diastereomers in all solvent. Potentially useful information could be contained in the chemical shift of CH, methyl and aromatic protons, which are somewhat different in the two complexes. However, this information is lost because of the local dynamics affecting this portion of the molecule and again no safe conclusion can be drawn from the averaged values observed.

The data in table 3.6 show that minor structural adjustments must occur between crystal and solution geometry, whichever of the two structural types is correct. The data suggest a counter-clockwise rotation of about 10° of the acetate arm with respect to the solid state structure for $\Lambda(\delta\delta\delta\delta)$, and the inversion of the macrocycle for $\Lambda(\lambda\lambda\lambda\lambda)$ (or their mirror images).

An empirical correlation between the ^{1}H-NMR spectral width of the Yb complex, that is the value of the magnetic anisotropy factor, $\mathcal{D}$, and the preferred conformation has been proposed [39, 69]. With reference to the major and minor forms of YbDOTA [12, 18], it has been assumed that a value of $\mathcal{D} \simeq 5000$ in water or in methanol solution may be associated with a *n*-type structure of the ytterbium complex, while a $\mathcal{D} \simeq 3500$ is evidence for the *p*-type molecule.

Unfortunately, the reference molecule binds the lanthanide ions with different donor atoms: a comparison between the coordination properties of a carboxylate oxygen and an amide oxygen is neither straightforward nor safe. Moreover, such complexes are involved in different hydration-dehydration equilibria [18], thus the magnitude of the magnetic anisotropy is likely to be largely determined (beside the specific contribution of the octadentate cage) by the presence or absence of an axial substituent.

According to Bleaney's model [64], $\mathcal{D}$ is directly dependent on the crystal field experienced by the lanthanide ion, which is in turn strongly affected by a ninth capping ligand [45, 46, 70].

The observation of a certain regularity in the anisotropy factor, $\mathcal{D}$, is demonstrated by recording spectra of each complex in DMSO. Such a solvent has a large effect on the value of $\mathcal{D}$: the spectra of both complexes, spanning about 170 ppm in MeOD, water or acetonitrile, reduce to 100 ppm in DMSO solution, with axial A_1 and acetate C_2 protons resonating at about 70 ppm and -35 ppm, respectively. In spite of this difference, the pseudo-contact shifts in DMSO are perfectly proportional to the corresponding values in CD_3OD and can be fitted smoothly to the two sets of geometrical factors derived from the n- and p-type structures (see Table 3.6). This proves that the complex possesses the same coordination, but again it is impossible to rule out either of the two possibilities for the helicity of the ring.

The observed variation of ^{1}H-NMR spectral width must be connected to some variation in the properties of the central Yb cation. Furthermore, whether or not the conformation of the macrocycle has switched between λ/δ as a consequence of the solvent variation must be determined, as well.

NOE effects

Unequivocal evidence for the p or n conformation of the complex must be found from an independent measurement. As demonstrated in the case of YbDOTMA, information can be obtained by observing the steady-state NOE spectra following saturation of relevant ring or acetate protons. In this way, it is possible to estimate some pairwise proton distances and constrain the geometry of the complex to one of the two diastereomers.

With these complexes, this experiment is not straightforward, owing to:

- the large self-relaxation rates ρ induced by the coupling to the unpaired electron, which sizeably reduce the Overhauser enhancement η;

- the dependence of σ on the correlation time in magnitude and sign:

$$\sigma \simeq \frac{2}{3}J^2 - \frac{1}{3}J^0 = \frac{2\tau_c}{1 - 4\omega^2\tau_c^2} - \frac{\tau_c}{3}$$

At 300 MHz, such a contribution is expected to vanish for $\tau_c \simeq 260$ ps, for which no NOE can be observed [65, 71].

On saturating the C_2 resonance, a clear NOE can be obtained on the A_2 proton; furthermore, on saturating the C_1 peak, a NOE is obtained on E_2. The latter effect cannot be justified for a twisted antiprism (p-type structure), highlighted by the interproton distances reported in Table 3.7 [7]. The $\Lambda(\delta\delta\delta\delta)/\Delta(\lambda\lambda\lambda\lambda)$ form can be assigned to the Yb·(R)-**3** complex in solution, consistent with that observed for the crystal structure and different to the previous suggestion [36, 39] based only on the magnitude of the anisotropy $\mathcal{D}$.

Eu·(R)-**3**			Na·(S)-**4**		
Proton	C_1	C_2	Proton	C_1	C_2
A_1	3.37	3.99	A_1	3.45	3.47
E_2	**2.19**	3.35	E_2	2.44	2.62
A_2	2.92	**2.12**	A_2	2.52	**2.09**
E_1	2.72	2.84	E_1	2.51	3.47

Table 3.7: Mean interproton distances (Å) obtained from the crystal structures of Yb·(R)-**3** (square antiprism, n-type structure) and Na·(R)-**4** (distorted square antiprism, p-type structure). The values which allow the detection of a steady-state NOE are typed in boldface.

The case of Yb·(S)-**4** is more intriguing. On saturating each of the ring protons, no NOE effect can be clearly observed on the acetate peaks nor even between geminal protons: this demonstrates that the correlation time of this complex is in the intermediate-motion region where $\sigma = 0$. In order to get over the problem, an ROE effect, for which the zero value never occurs, can be sought instead. It is, anyway, very difficult to effectively spin lock widely separated signals such as those under examination [65].

It was found more convenient to act on the correlation time of the molecule, making it longer, by changing the solvent from MeOD to the more viscous solvent DMSO in which the re-orientational tumbling of the complex slows down.

A further issue has to be checked at this point: that the structure of the two complexes is independent of solvent composition. Such a conservation was verified on the phenyl derivative repeating the NOE experiments performed in CD_3OD: the same results were obtained in DMSO and acetonitrile. Furthermore, on changing the temperature, the absence of observable dynamic processes involving the DOTA-portion of the molecule was also demonstrated in the other solvents.

In the case of Yb·(S)-**4** in DMSO, upon saturating the C_1 resonance at -16.2 ppm, an NOE was revealed on the E_2 peak at 12.0 ppm. Such an effect is

---

[7]From the signal-to-noise ratio of the Overhauser effect between the two geminal acetate protons after 32 K acquisitions, the limiting distance between pair of protons allowing the relevability of a NOE can be assessed to about 2.4 Å(assuming comparable self relaxation rates).

distinct and negative, since the slow-motion regime has been reached, and proves that the conformation of the complex is of n-form, as found for the phenyl analogue. The configuration of the complex in solution is therefore $\Lambda(\delta\delta\delta\delta)$ or $\Delta(\lambda\lambda\lambda\lambda)$; it will be demonstrated that the former is the case for an R configuration at carbon, the latter for S.

3.4.3 YbDOTAM-Phe, YbDOTAM-Ile and YbDOTAM-Pro

The ^{1}H-NMR spectra of Yb·(S)-**5**, Yb·(S)-**6** and Yb·(S)-**7** in D_2O solutions were measured (see figure 3.14). They exhibit one main species in solution, whose signals are practically equivalent to those of the two amide complexes Yb·(R)-**3** and Yb·(S)-**4**.

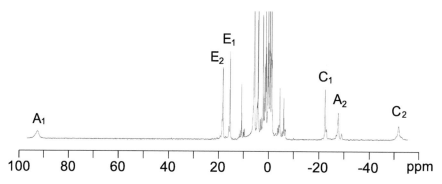

Figure 3.14: ^{1}H-NMR spectrum of Yb-DOTAM-Phe (Yb·(S)-**5**) in D_2O.

It must be noticed, anyway, that at room temperature a set of small peaks (integral < 5%) surrounds each line of the main species. Such phenomenon might be explained taking into account the presence of slow conformational equilibria involving the side arm portions not involved in the coordination or the possibility of a partial partecipation of the side arm ester groups in coordinating the ion.

3.4.4 YbTHP

Analysis of the pseudocontact shifts

The ^{1}H-NMR spectra of Yb-(S)-THP (Yb·(S)-**8**) was measured at pD 7 and is displayed in Figure 3.15(a). All the lines are unusually broad (>300 Hz) compared to analogous stable and non-labile ytterbium complexes [14, 27, 40], suggesting that in solution two or more species may be in intermediate exchange. The nature of such equilibrium, however, cannot be easily assessed owing to the large number of degrees of freedom of the systems (ring and side arm conformations, solvent coordination, oxydryl deprotonation,...).

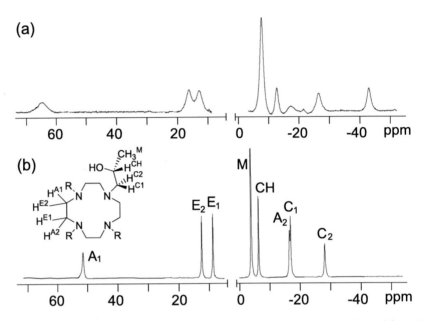

Figure 3.15: ^{1}H-NMR spectrum of Yb-(S)-THP in D$_2$O at pD=7 (a) and pD=2.5 (b). The proton labelling is also displayed.

On lowering the pH, the resonances progressively sharpen, showing that the main dynamic process is associated with a deprotonation of one of the bound hydroxyalkyl groups or of a hypothetically bound water. At pD 2.5, only the most shifted lines feature a width larger than 100 Hz (Figure 3.15(b)): such a result is compatible with only one species in solution [14, 72] and allows the assignment and the full interpretation of the spectral pattern. Eight significantly paramagnetically-shifted resonances can be detected, arising from each of the chemically inequivalent protons of the complex, which is now not involved in any observable dynamics over a wide temperature range (0 to +60°C): the lines showed no evidence of exchange broadening and their position was only affected by the temperature dependence of the ytterbium-induced paramagnetic contribution [65]. The assignment of the resonances was made from the relative peak areas, by comparison with the spectral data previously reported for YbDOTA [14]; a 2D correlation spectrum (COSY) was necessary to discern between the high field lines (C_1 and C_2, CH and A_2 protons, according to the labelling of figure 3.15). The shift values for each assigned signal are summarised in table 3.8.

A set of reference geometrical factors for structural optimization is usually calculated from crystal structures [14, 40, 72].

As previously discussed, the X-ray data available on the stereomer mixture of Eu-THP can provide geometrical parameters corresponding to both p- and n-type structures, since two enantiomeric pairs co-crystallize from solution [22]. The homochiral $(SSSS)$ molecule is endowed with a distorted antiprismatic geometry $(\Delta(\lambda\lambda\lambda\lambda))$, while a reference geometry corresponding to a square prismatic conformation can be derived from the $(SSSR)$ structure, once the configuration of the fourth chiral centre is inverted. The values of the distances between ytterbium and the donor atoms on the ligand deserve further attention. Indeed, in cases of homotypical series on lanthanide complexes, the Yb-N e Yb-O distances are 0.08-0.1Å shorter than in the related Eu complexes [39]. Thus, a MM2 calculation was run on each reference structure with constraints on the coordination distances, shrinking the Yb-O and Yb-N distances to 2.3Å and 2.6Å, respectively.

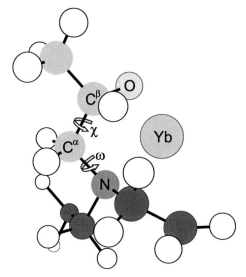

Figure 3.16: Dihedral angles ω and χ in the side arms used in the optimization procedure.

The lower symmetry of the crystal structure was compensated by such optimization, so all the equivalent protons of the molecule featured the same geometric factors. The two sets obtained this way for the two forms are displayed in Table 3.8.

Two main degrees of freedom were considered in the molecule: the torsion angle ω around N and C^α (defined here as the dihedral Yb-N-C^α-C^β shown in Figure 3.16) and the torsion angle χ around C^α and C^β (defined here as the dihedral N-C^α-C^β-Me shown in Figure 3.16). Thus a simultaneous fitting of ω, χ and the magnetic susceptibility $\mathcal{D}$ was performed with reference to eq. 3.2, starting from the two geometries for the p- and n-forms, as summarised in Table 3.8.

proton	$\left\langle \dfrac{3\cos^2\theta - 1}{r^3} \right\rangle \times 100$		D$_2$O			DMSO		
	p-form	n-form	δ^{exp}	$\delta_{pc}^{exp} - \delta_{pc}^{calc}$ p-form	$\delta_{pc}^{exp} - \delta_{pc}^{calc}$ n-form	δ^{exp}	$\delta_{pc}^{exp} - \delta_{pc}^{calc}$ p-form	$\delta_{pc}^{exp} - \delta_{pc}^{calc}$ n-form
A_1	3.00	3.15	47.8	1.5	-0.3	46.0	1.3	-0.4
A_2	-1.54	-1.50	-21.2	-1.5	-0.1	-17.8	0.0	1.3
E_1	0.40	0.37	9.7	0.1	1.0	10.1	0.1	1.0
E_2	0.48	0.50	6.0	-2.5	-0.9	6.8	-2.1	-0.6
C_1	-0.92	-0.91	-9.4	0.4	2.9	-7.8	1.5	3.0
C_2	-2.34	-1.92	-31.0	0.4	-3.7	-28.3	0.5	-3.3
CH	-1.77	-1.64	-20.5	2.6	2.7	-20.2	0.8	0.9
M	-0.42	-0.34	-5.4	-1.9	-1.6	-4.6	-2.2	-1.9
$\langle r \rangle$ (ppm)				1.4	1.5		1.1	1.5
R %				7.0	8.8		6.0	8.4
$\mathcal{D}$ (ppm Å^3)				1460±60	1490±70		1380±40	1410±50

Table 3.8: Experimental lanthanide induced shifts (δ^{exp}, in ppm) of the Yb-(S)-THP complex in D$_2$O and d$_6$-DMSO solutions (T=25°C) and differences between the experimental pseudocontact contributions and those calculated in the fitting algorithm. The proton labelling is analogous to that of figure 3.15. The factors on the second and third columns are calculated from the model geometry as described in the text and refer to structures which are representative of a n- type and a p-type, respectively. The experimental pseudocontact shifts used in the computations are the values of the diamagnetic lutetium complex [22]. For each set of chemical shift fitted, the mean value (r) of the difference $\delta_{pc}^{exp} - \delta_{pc}^{calc}$, the agreement factor R and the susceptibility factor $\mathcal{D}$ are displayed.

The side arm was thus systematically rotated around the two angles ω and χ to obtain the best agreement between the geometrical factors and the observed pseudocontact shifts. Only minor structural adjustments were required to fit the solution data, whichever of the two structural type is correct: in both cases, the optimal variations of the two torsion angles was less than 5°, with no significant alteration in the Yb-hydroxyl distance.

First of all, it must be noticed that such results correspond to a coordination geometry of Δ type and confirm the nature of the chirality amplification depicted by the crystal data [22], defining the local stereochemistry around the metal ion. Indeed, the fact that the stereogenic centre belongs to the inner sphere, acts in the optimisation algorithm as an experimental configurational correlation, as was the case for chiral YbDOTMA [40].

Secondly, as far as the ring is concerned, we must recall that in the case of carboxylate and amide lanthanide complexes, ^{1}H-NMR shift data could not simply discern between isomeric forms and only by NOE measurements could one assign which was the more stable structure in solution [40, 72].

The situation seems to be reversed here. We can indeed observe that the two sets of parameters are not perfectly proportional: on going from the p- to the n-type structure, while the ring protons exchange between pairwise equal positions, the nuclei on the side arms (mainly: the methyl and CH protons) experience a different dipolar contribution. As reported in Table 3.8, the geometrical parameters derived from a p-type structure allow a slightly better fit of the experimental chemical shifts, with an agreement factor somewhat smaller than in the case of the other diastereomeric form.

In this case, however, while the 1D ^{1}H-NMR spectrum seems to yield a more complete structural refinement, we found it impossible to prove such conclusion by an Overhauser effect. From the signal-to-noise ratio of the Overhauser effect between the two geminal acetate protons after 32 K acquisitions, the limiting distance between pair of protons allowing the relevability of a NOE can be assessed to about 2.4 Å(assuming comparable self-relaxation rates) [65, 72].

As shown in Table 3.9, a n-type structure ($\lambda\lambda\lambda\lambda$ conformation of the cyclen) can be excluded only on the basis of the absence of the NOE between E_2 and C_1, while no positive evidence can be found to support the p-type conformation. What is more, this experiment is not always simple on paramagnetic species, owing to the large self-relaxation rates induced by the coupling to the unpaired electron and to a correlation time for these chelates often close to the critical value of 260 ps, where no NOE can be observed at 300 MHz [65].

The simultaneous observation of the A_2-C_2 enhancement and lack of E_2-C_1 (which should be more easily detected owing to more favorable T_1 and T_2 of both protons) can be cautiously taken as a proof of the p-type structure in solution, which could be fully described as $\Delta(\lambda\lambda\lambda\lambda)$ (figure 3.17).

It must be stressed that such a finding would indicate the persistence of the conformation of the macrocycle with respect to the crystal data of the homochiral Eu analogue.

Square antiprism (n-form)				Distorted antiprism (p-form)		
Proton	C_1	C_2		Proton	C_1	C_2
A_1	3.56	4.15		A_1	3.62	3.64
E_2	**2.26**	3.53		E_2	2.42	2.69
A_2	2.99	**2.07**		A_2	2.60	**2.06**
E_1	2.80	2.92		E_1	2.56	3.63

Table 3.9: Distances between relevant proton pairs in the two model structures for the n- and p-type diastereomers of Yb-(S)-THP.

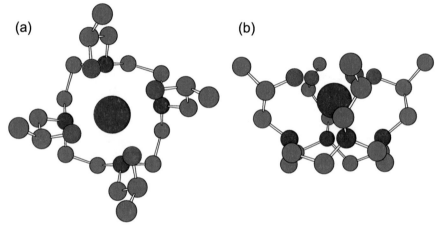

Figure 3.17: Top (a) and side (b) views of the solution structure of Yb-(S)-THP, as determined by the ^{1}H-NMR analysis.

3.4.5 [YbTHP]$_2$

ESI-MS spectra were recorded on a MeCN solution of Yb-(S)-THP with (figure 3.18(a)) and without (figure 3.18(b)) a base added.

In the latter case, the presence of a di-positively charged dimer is unequivocally proved: the molecular peak at 576 exhibits an isotopic pattern compatible with the presence of two ytterbium nuclei and a mono-charged molecular ion can be detected at twice the molecular weight. A second double deprotonation has taken place in these aprotic solvents and a dimerization has paired the doubly-deprotonated, mono-charged species.

Electronic spectroscopies will allow us to furtherly prove this phenomenon and to characterize it in deeper details (see section 3.5.4). Here the structural description of dimeric moiety is given on the basis of the ^{1}H-NMR data collected.

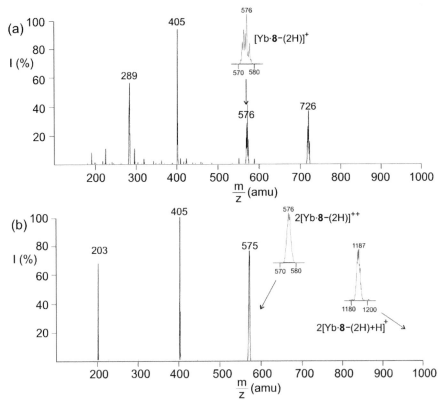

Figure 3.18: ESI-MS spectra of Yb-(S)-THP in MeCN without (a) and with (b) a base added.

Analysis of the pseudocontact shift

A ^{1}H-NMR analysis allows one to fully determine the structure of the dimer. After the addition of a base (Et$_3$N or even an excess of the free ligand) to a MeCN or DMSO solution, the resonances of the monomer progressively disappear and a new set of eight lines grows up (Figure 3.19).

These signals are particularly sharp ($\Delta\nu < 70 Hz$) and cover a spectral range 15 ppm wider than that of the monomer; they are compatible with the presence of a single, D_4 symmetric, species. The assignment, performed from the relative peak areas and by means of a 2D COSY, is reported in Figure 3.19 and the values of the shifts are displayed in Table 3.10. A new broad peak, integrating one half of the others, can be detected at around 120 ppm and can be assigned to the hydroxyls.

An optimisation similar to that performed on the monomeric species was then attempted. Since two ytterbium ions are now present in the molecules, the pseudocontact chemical shift experienced by each proton is the sum of the separated contribution from the different paramagnetic species (see figure 3.20),

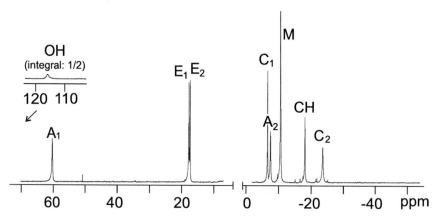

Figure 3.19: ^{1}H-NMR of dimeric Yb-(S)-THP (d$_3$-MeCN; c = 80 mM).

provided the magnitude of the coupling (if it occurs) between the two unpaired electrons is moderate ($J \ll kT$) [65].

The value to be optimised in the structural refinement is then:

$$\delta^{pc} = \mathcal{D} \left(\frac{3cos^2\theta_1}{R_1^3} + \frac{3cos^2\theta_2}{R_2^3} \right) + \Delta \qquad (3.4)$$

where (θ_1, r_1) and (θ_1, r_1) indicate proton positions with respect to the two Yb ions in the dimer. Given the symmetry of the complex, a consistent optimisation may be accomplished acting on one monomer only, adding to the three degrees of freedom previously considered (ω, χ, $\mathcal{D}$) the position of a second ion, free to move along the four-fold symmetry axis of the system; this is equivalent to optimizing the distance between the two monomeric units. The results of this procedure, performed on the two different p- and n-type reference geometries, are summarized in table 3.10. Again, the p-type structure is the one which fits better the experimental results; the presence of a second paramagnetic ion allows an even cleaner discrimination between the two structural types.

Neither a ring inversion nor side arm helicity interchange have taken place in the double deprotonation processes and the structure of the monomer in the dimeric complex can be considered unaltered. The distance between the two ytterbium ions is 4.2 ± 0.1 Å and is fully compatible with the formation of hydrogen bridges between the hydroxyls involved in the coordination (see Figure 3.21). The availability of bound hydroxyl moieties to hydrogen bond anionic molecules had already been stressed in the case of Cd^{2+} complexes with strictly analogue ligands, where a cavity above the oxygens plane acted as a molecular receptor for p-toluensulphonate and p-nitrophenolate [9].

A further geometrical parameter of the dimer is the twist angle between the two units, defined for example as the dihedral O$_1$–Yb$_1$–Yb$_2$–O$_2$. From the calculations described above, such parameter remains undetermined, since the pseudocontact contribution of the ytterbium ions expressed in eq. 3.4 is indeed cylindrically symmetric and no information about the dihedral distorsion along the Yb–Yb axis can be drawn from it. As the two monomers in the dimer are progressively rotated with respect to each other, the O-O distances d vary in a range between 1.8 Å and 2.6 Å. In order to have H-bridges, the O-O distance d must be constrained to $\simeq$ 2.3 Å, which allows one to choose the most twisted geometries. The sense of superhelicity, however, remains undetermined, and the two forms of Figure 3.22 are both compatible with the available NMR data.

Relaxation properties

If dipolar relaxation is the most efficient mechanism, for molecules in the fast motion regime, the relaxation rates are expressed by the equations 2.44 and 2.56.

proton	δ^{exp}	$\delta_{pc}^{exp} - \delta_{pc}^{calc}$ p-form	$\delta_{pc}^{exp} - \delta_{pc}^{calc}$ n-form
A_1	57.3	0.0	-1.6
A_2	-11.5	1.7	1.2
E_1	15.2	0.9	3.2
E_2	14.8	-0.4	0.6
C_1	-9.8	-1.8	-2.5
C_2	-26.8	0.4	-0.2
CH	-23.1	-1.6	-1.6
M	-12.4	0.8	0.8
$\langle r \rangle$ (ppm)		0.9	1.5
R %		4.4	6.7
$\mathcal{D}$ (ppm Å^3)		1580 $\pm$ 40	1610 $\pm$ 60
d (Å)		3.8 $\pm$ 0.2	4.0 $\pm$ 0.3

Table 3.10: Experimental lanthanide induced shifts (δ^{exp}, in ppm) of the Yb·(S)-THP dimeric complex in d$_6$-DMSO solution (T=25°C) and differences between the experimental pseudocontact contributions and those calculated in the fitting algorithm, as described in the text, with reference to structures which are representative of a p-type and a n-type, respectively. The experimental pseudocontact shifts used in the computations have been calculated subtracting the corresponding values of the diamagnetic lutetium complex [22]. For each set of chemical shifts fitted, the mean value $\langle r \rangle$ of the difference $\delta_{pc}^{exp} - \delta_{pc}^{calc}$, the agreement factor R, the susceptibility factor $\mathcal{D}$ and the distance d between the two ytterbium ions are displayed.

Measured longitudinal relaxation rates R_1 and reciprocal linewidths $\pi \cdot R_2$ for the monomer and dimer complexes are reported in Table 3.11.

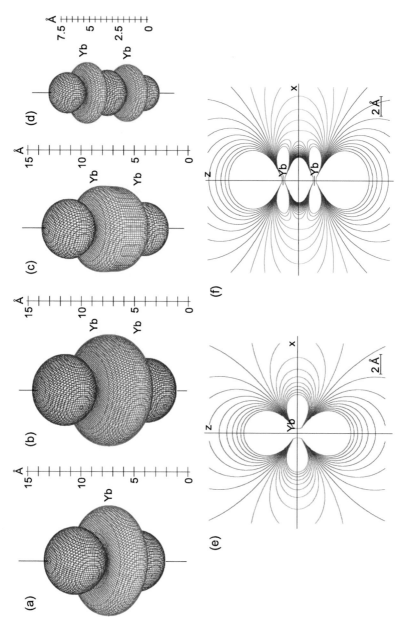

Figure 3.20: Spatial dependence of the pseudocontact shift in the monomeric and dimeric YbTHP, shown as surfaces of absolute constant values of δ^{pc}. Regions with positive δ^{pc} are represented in blue, those with negative values in red. (a) $\delta^{pc} = \pm 5$ ppm for YbTHP; (b), (c) and (d) $\delta^{pc} = \pm 5, \pm 10, \pm 100$ ppm for Yb[THP]$_2$; (e) and (f) represent iso-pseudocontact shift lines (from ± 10 to ± 100 ppm) in a section parallel to the C$_4$ axis in the two molecules.

Since the geometries of the molecules have already been determined accurately by means of the pseudocontact shifts, we can now use such structures to get deeper insight into some spectral features of these systems. In the case of the monomer, the experimental R_1 values correlate quite well with the $1/r^6$ values calculated from the model geometry (*corr.* $= 0.96$) according to equations 2.44 and 2.56. In the presence of chemical exchange the lines are broadened, preventing the use of reciprocal linewidths for structural inference. In fact, in the case of the R_2, the correlation is much poorer (*corr.* $= 0.38$).

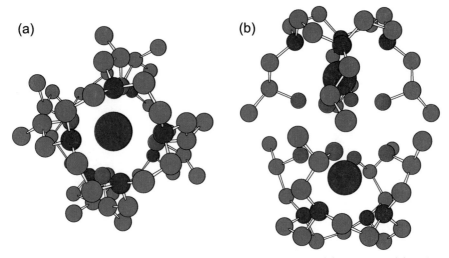

Figure 3.21: Optimized solution structure of dimeric Yb-(S)-THP; top (a) and side (b) views are displayed. The relative twist of the two monomers has been arbitrarily chosen.

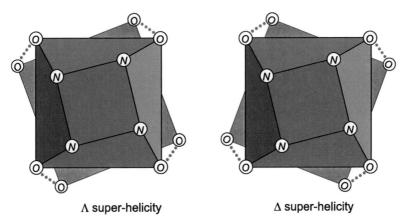

Λ super-helicity   Δ super-helicity

Figure 3.22: Possible Λ and Δ superhelicities deriving from the relative orientations of the two monomers in the dimeric complex.

	dimer						monomer			
	r(H-Yb$_1$) (Å)	r(H-Yb$_2$) (Å)	$\langle r \rangle$ (Å)	$1/\langle r \rangle^6$ ($10^{-4}\cdot$Å^{-6})	R_1 (s^{-1})	R_2 (s^{-1})	r(H-Yb) (Å)	$1/r^6$ ($10^{-4}\cdot$Å^{-6})	R_1 (s^{-1})	R_2 (s^{-1})
A_1	3.71	7.31	3.70	3.90	157	220	3.71	3.83	234	327
A_2	3.65	6.13	3.62	4.42	125	166	3.65	4.23	250	377
E_1	4.43	7.46	4.40	1.38	49	91	4.43	1.32	114	230
E_2	4.41	7.50	4.38	1.42	50	80	4.41	1.36	116	213
C_1	4.35	6.48	4.29	1.61	43	63	4.35	1.48	73	185
C_2	3.46	5.35	3.42	6.25	224	273	3.46	5.83	384	402
C_H	3.85	5.15	3.75	3.61	89	129	3.85	3.07	147	276
M			4.28	1.63	31	41		3.83	43	126
	CORREL.COEFFICIENT				0.95	0.93	CORREL. COEFFICIENT		0.96	0.38
	C ($\times 10^5$ s·Å^{-6})				3.2±0.1	4.3±0.2	C ($\times 10^5$ s·Å^{-6})		6.1±0.3	8±4

Table 3.11: Proton-ytterbium distances r in the two structures obtained for monomer and dimer Yb·(S)-THP. The measured longitudinal relaxation rates R_1 and reciprocal linewidths $\pi \cdot R_2$ for each signal were fitted against $\frac{1}{r^6}$ in the case of the monomer and $\frac{1}{\langle r \rangle^6} = \frac{1}{r_1^6} + \frac{1}{r_2^6}$ according to the relations of eq. 3.5. For each fitting, the correlation coefficient and the value of the proportionality constant C are reported.

In other words, even at pH 2.3, when only one species is apparently present in solution, dynamic processes still have place to some extent, due to a flexibility of the macrocyclic cage or to the (thermodynamically unfavoured, as we saw) axial exchange equilibrium. In the case of the dimer, on the other hand, both sets of experimental R_1 and R_2 values of the protons lead to rather good correlations with those calculated according to:

$$R_i = \sum_{n=1,2} \frac{C_n}{r_n^6} = \frac{C}{\langle r \rangle^6} = \frac{C}{r_1^6} + \frac{C}{r_2^6} \tag{3.5}$$

with the effective $\frac{1}{\langle r \rangle^6}$ values of the proton distances from the ytterbium ions taking the place of the usual $\frac{1}{r^6}$ summation.

Such results prove that the formation of the hydrogen bridges inhibits any possible dynamics concerning the axial substituent and the cage. Moreover, it is interesting to point out that the C value fitted in this second case for R_1 is almost one half of the corresponding figure obtained for the monomer.

3.5 NIR-CD analysis: chirality of the coordination cage

3.5.1 YbDOTMA

There are very few reports of CD and CPL of ytterbium in the literature [36, 73, 74] and no interpretation of the spectra was attempted, because no accurate information on the stereochemistry in solution was available. The present study takes as objective a molecule which is now very well structurally characterized. The presence of four homochiral methylacetate branches dictates the absolute stereochemistry of the coordination polyhedron, which is reflected in the signs and amplitudes of the dichroic bands, establishing a clear relationship between oberved CD and structure in solution.

An investigation of the electronic properties of lanthanide complexes of such a molecule was reported, [27] but interest was centered almost exclusively on the Eu(III) and Tb(III) complexes and a correlation of the observed spectra with plausible structures was impossible because of the complex nature of the electronic levels of such ions and the lacking of characterization of the solution dynamics of the species.

Results and discussion

The absorption and CD spectrum of Yb DOTMA in water is shown in figure 3.23.

The circular dichroism is readily recorded owing to the very favorable dissymmetry g-factor (up to about 0.25 at 946 nm). In fact the transition is

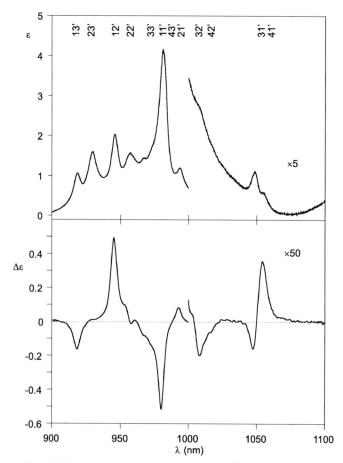

Figure 3.23: NIR absorption (upper curve) and CD (lower curve) spectra of YbDOTMA in H_2O ($l = 1$cm, $c = 0.16$ M). The second part of each spectrum (1000-1100 nm region) has been expanded by a factor of 5 and 50, respectively. On the top, each line is labelled according to the assignment proposed in the text (see figure 4): the numbers 1,2,3,4 and 1',2',3' refer to the energy levels of the $^2F_{7/2}$ and $^2F_{5/2}$ states of Yb(III) in the major form of the complex.

magnetically allowed and it must just borrow some electric dipole moment (e.g. from a *4f-5d* transition) to give rise to non-vanishing rotational strength. One can recognize several well-resolved components, between 1045 and 920 nm, with linewidth below 10 nm; they are more easily distingushed in the CD spectrum owing to sign alternation.

The signs and intensities of the sequence of bands apparent from the CD spectrum is related to the dissymmetric distribution of the atoms belonging to the organic ligand and in particular the four oxygen and four nitrogen donor atoms.

Relative contribution of the two forms to the CD spectrum

The existence of two species in solution raises the problem of determining their allied contribution to the CD spectrum, wheighted for their molar fraction. In principle, the two forms might contribute with largely different rotational strengths, to the point that *a priori* one might not neglect the presence of the minor component, even if its molar fraction never exceeds 10%. Moreover, one has no indication of the locations of the transitions of the p-form, which might or not be degenerate with those of the n-form, since the two have appreciably different coordination polyhedra. One can take advantage of the reversal populations of the two forms between Yb DOTA and DOTMA. Indeed, the p structures of these two complexes are similar, as are the n-forms, but what is the minor component for Yb DOTA is major for Yb DOTMA and vice versa, thus Yb DOTA can be reasonably assumed as a model for the minor form of Yb DOTMA. Naturally, being DOTA achiral, one must resort to comparing the NIR-absorption spectra, shown in figure 3.24.

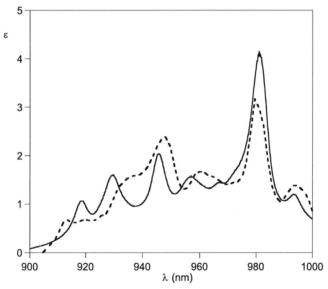

Figure 3.24: NIR absorption spectra (900 – 1000 nm region) of YbDOTA (dashed line) and YbDOTMA (solid line) in (l=1 cm, c= 0.01 and 0.162 M, respectively).

Although the individual intensities are somewhat different, the positions of all the bands are very similar. This leads to the conclusion that also the dichroic bands of the two forms of Yb DOTMA must be almost degenerate and that the spectrum of the n-type form has to be looked for in the same spectral region. To demonstrate that the most prominent features of the CD spectrum can be ascribed to the more abundant species only, one can displace the equilibrium position, by changing solvent composition and temperature. In particular, upon addition of a large excess of KF to the acqueous solution, the minor (n) form becomes much less populated (by ^{1}H-NMR it is evaluated about 2.5 %), once more in reverse analogy with DOTA. The CD spectrum of this sample is superimposable to that without fluoride, which is a positive indication that it can be safely attributed essentially to the major form. The only sizable effect of the excess of KF is a slight enlargement of the whole CD pattern, which can be related to a dependence of the crystal field parameters on the ionic strenght of the solution. Moreover, also the spectrum in methanol (where the solution compostion is comparatively more favorable to the minor form) is nearly identical to that in water. From this point on, we shall assume that the minor species contributes to the main features of the CD spectrum to an extent not significantly larger than its molar fraction (i.e. that the rotational strengths of the two forms are comparable). It is noteworthy that in MeOD the counterion is partly axially bound [40], but this does not apparently affect the electronic spectrum.

Variable-temperature spectra

In a C_4 symmetry, the ground and excited states, $^2F_{7/2}$ and $^2F_{5/2}$ split into 4 and 3 doubly degenerate sublevels. Owing to the reduced spectral width (about 120 nm), it is likely that the splitting of the $^2F_{7/2}$ is comparable to $k_B T$ (≈ 2.5 kJ mol^{-1} at 298 K) and that several levels are populated at room temperature. Indeed, one must observe that both the absorption and CD spectra are rather complex and show more than just three components (those originating from the lowest-lying level of the ground state).

In order to gain insight into the individual transitions of the multiplet and their rotational strengths, low-temperature spectra were recorded. At -80° C in methanol, all the lines become narrower and appear consequently more intense, as shown in figure 3.25

The position of the whole set of lines is only slightly affected and such an effect can be safely explained with a structural stiffness and with a minor contribution of less defined forms as the temperature goes down.

Among the transitions observed in the CD spectrum, only four were sufficiently well-resolved to allow for a quantitative determination of rotational strengths. Rotational strenghts for these resolved transitions can be determined by integrating the observed CD in the wave number domain over transition profiles and then evaluating [75]:

$$R_{ij}(T) = 0.248 g_i \int_{i \to j} \frac{\Delta \epsilon(\tilde{\nu})}{\tilde{\nu}} d\tilde{\nu} \qquad (3.6)$$

where R_{ij}(T) denotes the rotational strenght of a transition i→j, expressed in units of D^2 (D=1 Debye unit $= 10^{-18}$ esu cm); g_i is the electronic degeneracy of the level i ($g_i =2$ in the present case) and the integration is over the CD profile of the i→j transition. Over a Lorentzian lineshape of $\Delta \epsilon^{max}$ amplitude

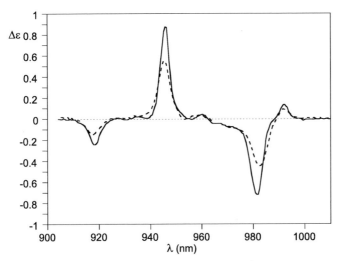

Figure 3.25: Variable temperature NIR-CD spectra (900-1020 nm region) of YbDOTMA in MeOH (l=1cm, c=0.096M): 193 K (solid line) and 298 K (dashed line).

and $2\Delta\tilde{\nu}_i$ half-height linewidth, the integration of eq. 3.6 is given by:

$$R_{ij}(T) = 0.124g_i(\pi\Delta\epsilon^{max}\Delta\tilde{\nu}) \qquad (3.7)$$

To this end the spectrum can be deconvoluted to a sum of Lorentian lineshapes, affording the data of table 3.12.

λ (nm)	trans.	T=298 K			T=193 K		
		$\Delta\epsilon$ (±0.01)	$R(T)$ (esu²cm2)	R' (esu²cm2)	$\Delta\epsilon$ (±0.01)	$R(T)$ (esu²cm2)	R' (esu²cm2)
918	$1 \rightarrow 3'$	-0.15	-8.2 ± 0.1	-13.4 ± 0.2	-0.23	-9.1 ± 0.1	-12.9 ± 0.1
946	$1 \rightarrow 2'$	0.55	21.4 ± 0.5	34.7 ± 0.8	0.87	26.7 ± 0.5	37.7 ± 0.7
980	$1 \rightarrow 1'$	-0.45	-28.3 ± 0.3	-46.0 ± 0.5	-0.71	-31.1 ± 0.3	-44.0 ± 0.4
993	$2 \rightarrow 3'$	0.087	4.4 ± 0.1	7.2 ± 0.2	0.13	4.7 ± 0.1	-6.7 ± 0.1

Table 3.12: Temperature dependence of the rotational strenght of the main electronic transitions of the major form of YbDOTMA in MeOH. The bands are labelled according to the assignment shown in figure 5. $R(T)$ is the observed rotational strenght, calculated according to eq. 3.7, R' is the corrisponding value corrected for the Boltzmann factor $b(T)$ according to eq. 3.8 (see table 3.13).

Comparing the temperature-dependent rotational strengths $R(T)$ of the different transitions, one observes that the three bands at 980, 946 and 918 nm increase, while the one at 993 nm becomes weaker. This is compatible with the fractional thermal populations $b(T)$ of the ground state sublevels at room and low temperature calculated in Table 3.13; as a proof of our assignment of the transitions, it must be noted that the corrected rotational strenght R':

$$R' = \frac{R(T)}{b(T)} \qquad (3.8)$$

for the four transitions is in good approximation temperature independent (table 3.12).

sub level	b (298 K)	b (193 K)
1	0.616	0.718
2	0.338	0.283
3	0.029	0.005
4	0.015	0.003

Table 3.13: Boltzmann populations $b(T)$ for the crystal-field levels of the $^2F_{7/2}$ state of Yb in YbDOTMA.

Tentative assignement of the electronic spectrum

The splittings of the excited-state levels can be calculated from the frequencies of the three transitions assigned to the lowest-lying level, as about 320 and 380 cm^{-1}. By taking into account all the transitions that can be recognized in the spectrum, one obtains the assignment depicted in the bottom of figure 3.23 and in the diagram of figure 3.26.

This assignment is consistent with the variable temperature data. It is noteworthy that one component, namely the $2 \to 1'$ transition, is cleary visible in the absorption spectrum but has no detectable CD counterpart.

Interpretation of the spectrum of Yb DOTMA in terms of ligand-field effects is made difficult by the rather low number of transitions that are clearly observed in the spectrum.

In a C_4 symmetry the crystal field Hamiltonian of equation 1.14 can be decomposed into the sum:

$$\mathcal{H}_{cf} = \sum_{k,q} B_q^k U_q^k \qquad (3.9)$$

where $U_q^{(k)}$ are intraconfigurational spherical tensor operators of rank k (k=2, 4, 6) and order q ($q = 0, \ldots, 4$ with $|q| \le k$) and the B_q^k are the corresponding crystal fiels parameters. The whole Hamiltonian is responsible for the position of the electronic transitions, as well as it participates in determining the intensity of isotropic and dichroic bands. [76]

The two terms C_4^4 and C_4^6 in particular lead to a mixing of the M_J components of the two levels, J=7/2 and J=5/2. The values of the corresponding parameters depend on the degree of distortion of the coordination polyhedron,

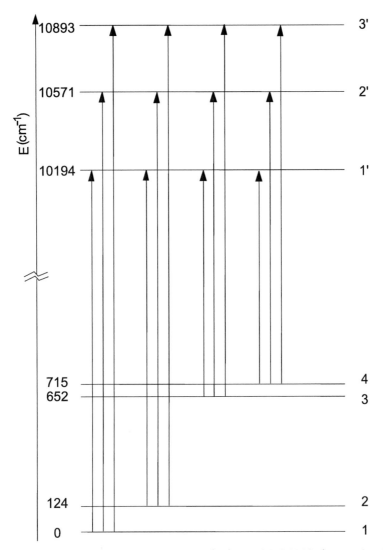

Figure 3.26: Electronic transition of Yb(III) in YbDOTMA (major form). The four lower levels (all doubly degenerate) belong to the $^2F_{7/2}$ state and the three upper ones to the $^2F_{5/2}$ state.

to the point that they are expected to vanish for a C_{4v} symmetry (cubic or antiprismatic) By comparison of the absorption spectra of Yb DOTA and Yb DOTMA, we observed that the transition frequencies are practically coincident. Considering that the two complexes differ in the layout of the atoms around the metal ion and in particular in the degree of distortion of the coordination polyhedron, it appears that the non–diagonal terms, B_4^4 and B_4^6 affect only to a minor extent the energy levels of both the ground and excited electronic states of Yb^{3+} in this complex. Another very relevant aspect emerges from this comparison, regarding the B_0^2 term. According to Bleaney, [64] it should be proportional to the magnetic susceptibility anisotropy, $\mathcal{D}$. This parameter is very different for Yb DOTA and Yb DOTMA, but once more the electronic transitions span the same spectral width. These remarks lead us to the conclusion that the comparison between optical and NMR spectra can be misleading.

3.5.2   YbDOPEA and YbDONEA

NIR-CD spectra in methanol solutions

The derivation of the distortion of the coordination polyhedron from the nature of the remote chiral centres is impossible, a priori, because of the absence of any clear relationship between the (known) chirality of the stereogenic carbons on the side substituents and the (unknown) chiralities of the ring and the layout of the pendant arms. The problem thus becomes a **configurational assignment** of the **absolute stereochemistry** about the metal ion.

Chiroptical data can provide the desired information. For this purpose, the electronic properties connected with the unfilled f shell of ytterbium can be exploited and a comparison with similar compounds of definite stereochemistry can be used for an unequivocal assignment. The near-infrared CD spectra of compounds Yb·(R)-**3** and Yb·(S)-**4** are shown in figure 3.27.

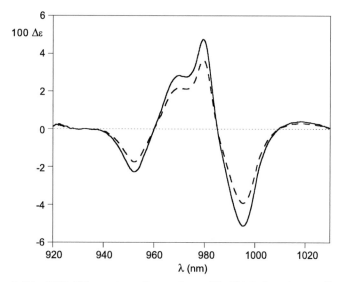

Figure 3.27: NIR-CD spectra of complexes Yb·(R)-**3** (continuous line) and Yb·(S)-**4** (dashed line) in MeOH solutions (c=58.6 mM and 83.1 mM; l=1 cm). To make the comparison easier, the curve for Yb·(S)-**4** has been inverted.

In the case of Yb·(S)-**3**, fluorescence emission spectra have been recently reported [37]; the results described herein constitute the absorption counterpart and are an important complement, being related to the ground state chirality.

Due to the large number of superimposing hot bands and to the linewidth of the signals, no detailed interpretation is feasible. The spectra of the two complexes are identical as far as the overall pattern is concerned (that is the band positions and relative intensities); the only difference is in the amplitude of the signals. This seems to suggest that the electronic properties of the central cation are dependent on the first coordination sphere only, whose distortion must in turn be insensitive to the replacement of the aromatic moiety. Such a conclusion is in agreement with that found in the fluorescent emission studies [37].

For each molecule, both in absorption and in CD, four main peaks were easily identified at 946, 975, 982 and 995 nm, with bandwidths of 10-20 nm; a fifth weak and broad signal was recognised at around 1020 nm. The dissymmetry factors calculated for the 995 nm band g_{abs}^{995} are 0.18 for Yb·(R)-**3** and slightly less (0.16) for Yb·(S)-**4**.

Comparison with YbDOTMA

It is instructive to compare the aforementioned results with the NIR-CD spectrum of (R)-YbDOTMA, which has been discussed in sction 3.5.1 and whose structure has been determined to be of $\Lambda(\lambda\lambda\lambda\lambda)$ type [41].

In the latter case, the coordinating oxygens of the ligand are charged carboxylates. The nature of the oxygen donor determines the magnitude of the crystal field splitting, believed to have a strong contribution from atomic charges. This can be the cause of the fact that the spectra of compounds Yb·(R)-**3** and Yb·(S)-**4**, which feature neutral amide carbonyls, cover a much narrower range (940-1020 nm compared to 915-1060 nm).

Notwithstanding the difference in spectral width, the similarity between the two sets of spectra, once the narrowing effect is taken into account, is striking. Also the relative amplitudes of the individual signals are comparable, with a g factor for the component at 980 nm in YbDOTMA calculated to be 0.2.

These features are sufficient to allow a safe correlation to be drawn for the stereochemistry of the metal centre. By combining the information obtained from NMR, the conformation of the two (R)- amide complexes can now be definitely described as $\Lambda(\delta\delta\delta\delta)$. Moreover, in solution as well, for each of these complexes, the same chirality of the substituents on the side arms, determines the same overall distortion of the coordination polyhedron ((R)$\rightarrow \Lambda$; (S) $\rightarrow \Delta$), which in turn is translated into the same sign sequence for the main CD bands in the NIR region.

Variable temperature data

In order to gain further insight into the composition of the CD pattern, a low temperature spectrum of Yb·(S)-**4** in methanol solution was recorded (figure 3.28).

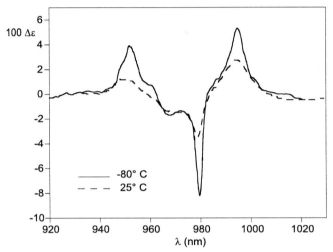

Figure 3.28: NIR-CD spectra of Yb·(S)-**4** in methanol at 25° (dashed line) and at -80°C (continuous line) (c=20mM; l=1cm).

At -80° C, the position of the signals does not change, but all the lines become narrower and, consequently, more intense. The effect is more pronounced for the band at 980 nm, which can thus probably be assigned to a transition originating from the ground crystal field sub-level of the $^2F_{7/2}$ state. Unfortunately, the narrowing of the other bands is not sufficient to allow a clear assignment of the crowded pattern, unlike YbDOTMA [41]. As a consequence of the weaker crystal field, not only do many signals fall into a narrower spectral window, but the hot bands remain significantly populated even at -80° C.

3.5.3   YbDOTAM-Phe, YbDOTAM-Ile and YbDOTAM-Pro

The same problem faced in the previous section holds for complexes Yb·(S)-**5**, Yb·(S)-**6** and Yb·(S)-**7**, for which the chirality transfer from the stereochemistry of the side arm to the distortion of the coordination cage could not be clarified by ^{1}H-NMR .

NIR-CD spectra of compounds Yb·(S)-**5**, Yb·(S)-**6** and Yb·(S)-**7** were recorded and are reported in figure 3.29. The three spectra are perfectly superimposable and feature the same four main bands at 946, 975, 982 and 995 nm of the analogue Yb·(R)-**3** and Yb·(S)-**4** discussed in the preceding section. Only the relative proportions of the two positive peaks at 975 and 982 nm are slightly changed.

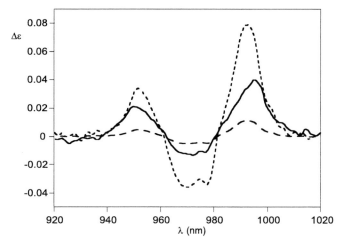

Figure 3.29: NIR-CD spectra of Yb·(S)-**5** (solid line), Yb·(S)-**6** (dashed line) and Yb·(S)-**7** (dotted line) in MeCN solutions (c=5 mM, 1 mM and 10 mM, respectively; l=1cm).

These results unambiguously prove that, here again, the distortion of the co-ordination polyhedron is of Δ type for the (S) isomers. At the same time, the perfect coincidence of the spectral pattern shows that the nature of the atoms outside the first coordination sphere of Yb^{3+} is hardly relevant in affecting the crystal field of the ion.

3.5.4   YbTHP

Low pH measurements

All the data collected via ^{1}H-NMR can be confirmed, completed and rational-ized by a study of the electronic properties of the central ytterbium cation.
The absorption and CD spectra of Yb-(S)-THP were recorded at pH=2.3 and are reported in figure 3.30 (continuous line).
The spectra of the tetraprotonated complex features three main well-resolved lines at 916, 941 and 970 nm; weaker transitions can be detected in the 950-960 nm region and at about 1005 nm. The dissymmetry factor is very high for the whole pattern, with a maximum value of 0.1 calculated at 940 nm.
It is instructive to compare the aforementioned results with the NIR-CD spec-trum of another C_4 symmetric ytterbium complex, Yb-(R)-DOTMA (figure 3.23) (see section 3.5.1 for the discussion) and whose structure is $\Lambda(\lambda\lambda\lambda\lambda)$ type.
The similarity between the two sets of spectra is striking. Also the amplitudes of the individual signals are comparable, with the component at 980 nm in

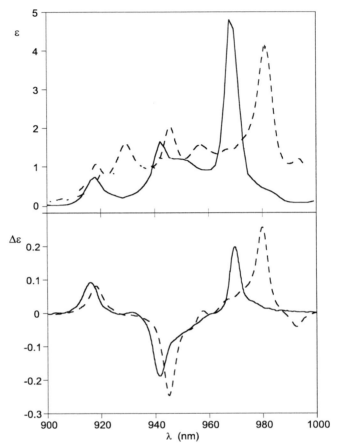

Figure 3.30: NIR absorption and CD spectra of Yb-(S)-THP (continuous line; H_2O, pH=2.5; c=20 mM; l=1 cm). The same spectra of Yb·(R)-DOTMA have been reported, inverted to make the comparison easier.

YbDOTMA having a $\Delta\epsilon$ of -0.2 and a g factor calculated to be 0.1.

In the latter case, the coordinating oxygens of the ligand are charged carboxylates instead of neutral hydroxyls; the nature of the oxygen donor contributes (via charges and polarizabilities) to the magnitude of the crystal field splitting: this can be the cause of the fact that the spectra of the two molecules cover a slightly different range, with an apparent correspondence between peaks shifted of 5-10 nm. These features are sufficient to draw a safe correlation for the stereochemistry of the metal centre. The chiral carbon on the side arm is rather close to the paramagnetic moiety and a configurational assignment of the absolute stereochemistry about the metal ion may be performed through the NMR analysis: the geometrical parameters of groups bound to the carbinolic carbon on the side arms are not compatible with a Λ distortion of the polyhedron. Chiroptical data confirm such information, describing the solution conformation of Yb-(S)-THP as $\Delta(\delta\delta\delta\delta)$.

Variable pH studies

Electronic spectroscopy is very efficient in characterizing the presence of exchanging species in solution, since its timescale is much faster than the times required by chemical equilibria and, as a consequence, the signal becomes the simple superposition of that of the different species. A whole titration performed in water in the pH range 2.3-10 was followed by NIR absorbance and CD (figure 3.31).

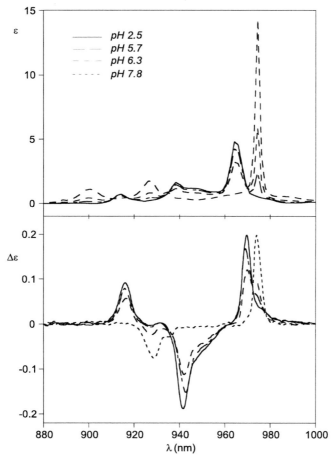

Figure 3.31: Variable pH NIR-CD spectra of Yb-(S)-THP (H_2O; c = 20 mM; l = 1 cm).

Starting at pH=2.3 and on raising the pH, both spectroscopies highlight the progressive disappearance of the species present in acidic conditions and the simultaneous formation of a second species. An analysis of the titration points is consistent with the presence of a first equilibrium with pK_a of about 6.5 ± 0.1 which leaves, at pH=7.8, a single, monodeprotonated molecule in solution. Two main absorption lines are now detectable, at 980 nm (positive CD) and 925 nm (negative CD), with a shoulder at about 935 nm. It must be noted that the dissymmetry factors of the transitions are comparable to those of the acidic species, even if the lines have moved and fall in a narrower spectral region. Nevertheless, the overall band sequence is compatible with a helicity of the coordination polyhedron analogue to the tetraprotonated complex.

On raising the pH up to 9, no significant variation could be seen in the spectrum, while, between pH 9 and 10, a progressive precipitation takes place, probably due to a second deprotonation followed by the formation of poly-hydroxyalkyl species [77, 78]. This datum is compatible with a second pK_a of about 9.5 ± 0.5, which is in agreement with the pK_a of 9.4 ± 0.1 published for the analogous Lu complex [47].

Studies in non-protic solvents: dimerization

The NIR absorbance and CD spectra of the complex dissolved in MeCN or DMSO are displayed in figure 3.32(a).

The presence of two forms in equilibrium can be deduced from the peak sequence, which is initially the same as the one obtained in water between pH 5 and 7 and related to a deprotonation equilibrium. After the addition of a base (Et_3N or even free THP ligand), the spectral pattern slowly evolves, with the contribution of the tetraprotonated form getting smaller. Nevertheless, unlike what happens in water, a mono-deprotonated species is not the result of such process and a third different spectrum can be recorded after 24 hours (figure 3.32).

The CD is dominated by two main contributions at 941 and 983 nm, with a shoulder at 936 nm and two weak bands at 915 and 930 nm. With respect to the starting pattern, the positions of the lines are only slightly changed, the $|g|$ factors of the transitions are much weaker and most notably their signs are inverted.

For structural purposes, anyway, it is not straightforward to assert that in this new species the central ion experiences an environment endowed with an inverted chirality. The empirical correlations used so far for lanthanide complexes are based on the comparison of CD spectra originating from rather similar chromophores: in the present species, however, the crystal field acting on the Yb ion is likely to be completely different with respect to the one described in the preceding section, as far both the charge distributions and the geometries around the metal are concerned; as a consequence, the electronic levels get shuffled in the *basification* process, so a detailed analysis of the whole pattern (deconvolution and assignment) should be necessary before extracting a piece of information from the CD datum.

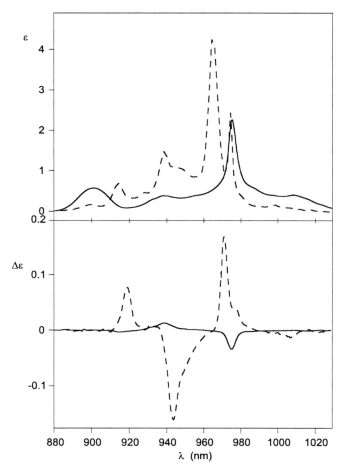

Figure 3.32: NIR absorption and CD spectra of Yb-(S)-THP in MeCN before (dashed line) and after (continuous line) a base was added (c = 80 mM; l = 1 cm).

3.6 Axial dynamics

3.6.1 YbDOPEA and YbDONEA

^{1}H-NMR data

The NOE experiments in different solvents unambiguously demonstrate that the conformation of the complexes is the same, independent of the solvent, and it is square anti-prismatic, as the crystal structure had suggested. The explanation of the change in the anisotropy term $\mathcal{D}$ in each case must imply that there is a change in axial coordination as the solvent is varied.

Useful information can be obtained by analysing the behaviour of the NMR spectra while changing the composition of the solvent. The addition of a different solvent causes a progressive variation of the position of the peaks, and hence of the associated apparent $\mathcal{D}$ parameter.

For example, starting from pure acetonitrile, adding less than 1% of water or DMSO is sufficient to obtain a spectrum that is very similar to that in the pure new solvent. Thus in CD_3CN/water or CD_3CN/DMSO mixtures, the magnetic anisotropy apparently continuously shifts from 3700 (starting value in CD_3CN) to 3000 (water) or 2100 (DMSO).

Moreover, in the latter case the growth of a new broad peak at 2 ppm appears. It has been shown[39] that in wet CD_3CN, Yb·(S)-**3** is axially bound to water, which resonates as an exchange-broadened signal to higher frequency of all other resonances. The appearance of the unbound water peak at 2 ppm suggests a competition between DMSO and acetonitrile, the latter being released upon addition of the former. Unfortunately, we were not able to observe the correspondent signal for the bound (non-deuteriated) DMSO, which we assume to be too broad to be detected.

It is noteworthy that this continuous change in the spectral width must be interpreted as a rapid equilibrium between two species with well-defined $\mathcal{D}$ values. Adding $1\mu l$ of DMSO to a 500 μl of CD_3CN solution, one observes very broad signals for all peaks, which separate at lower temperature (-40C) into two sets: a minor form with larger $\mathcal{D}_m$, and a major form with smaller $\mathcal{D}_M$.

Comparing the magnetic anisotropies at different temperature is not straightforward, since this parameter is strongly T-dependent. Nevertheless a reasonable scaling leads us to estimate the following values at 25° C:

$$\mathcal{D}_m \simeq 6500$$

$$\mathcal{D}_M \simeq 3000$$

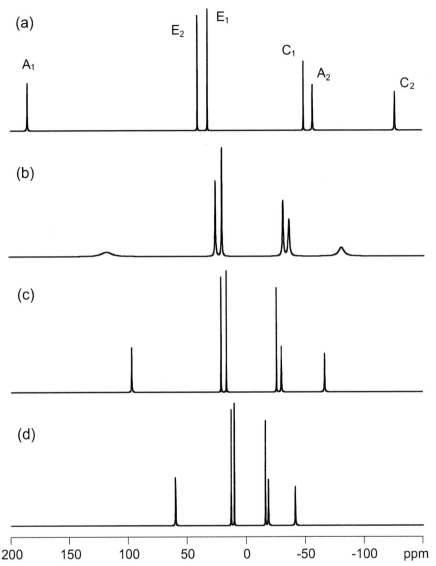

Figure 3.33: ^{1}H-NMR spectra of the DOTA-portion of Yb·(R)-**3**, simulated using the geometrical parameters of table 3.6 and different values of the $\mathcal{D}$ coefficient: (a) $\mathcal{D} \simeq 6500$ for the spectrum in dry acetonitrile; (b) $\mathcal{D} \simeq 3700$ for the spectrum in wet acetonitrile; (c) $\mathcal{D} \simeq 3000$ for the spectrum in water; (d) $\mathcal{D} \simeq 2100$ for the spectrum in DMSO. For the linewidths, values of T_2 relaxations were calculated from the NMR geometry, using equations 2.44 and 2.56, with $\tau_r \simeq 180$ ps and $\tau_s \simeq 0.2$ ps. The spectrum (b) has been obtained assuming dynamic interconversion from spectra (a) and (c), with 10-fold water eccess and a rate constant for water coordination determined by luminescence measurements as described in ref. [39].

Adding larger quantities of DMSO, the lines narrow and the minor form disappears. It seems reasonable to expect that value of $\mathcal{D}_m$ corresponds to that in pure acetonitrile, which is not the case. Our interpretation is that in this solvent, containing a non-negligible amount of water, there is an equilibrium between an anhydrous and a hydrated species, which is fast and leads to the observed shifts (figure 3.33).

Indeed, once more depending on the quantity of H_2O, one obtains different spectral widths and also different linewidths. At low temperature this solvent exchange process can be slowed down sufficiently so that the coordinated water may be directly observed (figure 3.34).

In this case the spectra were acquired at -40° C (10 mM Yb·(R)-**3** and 80 mM H_2O, 65.3 MHz) and the bound water resonates at +325 ppm , i.e. 321ppm to higher frequency of the free water resonance ($\Delta\nu = 21{,}000$ Hz). At -35° C the coordinated water signal broadens to such an extent that it may not be observed.

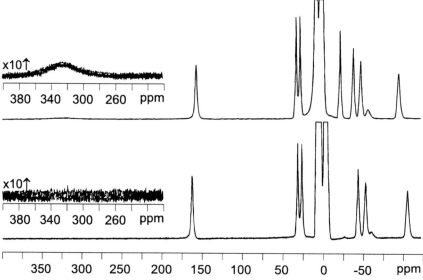

Figure 3.34: ^1H -NMR spectra of Yb·(R)-**3** (65.3MHz) at 233 K in CD_3OD (lower) and CD_3CN highlighting the bound water resonance at +325 ppm in CD_3CN. The small broad resonance at ca. -60ppm is an instrumental artefact. The amide NH resonance at -20 ppm has undergone H/D exchange in CD_3OD.

In a separate study, the ^1H-NMR spectrum of a 10 mM solution of dried Yb·(R)-**3** in dry CD_3CN was recorded and revealed a 4H singlet for the most shifted axial ring proton at +117.7ppm and an exchange broadened water resonance at +19 ppm (295 K, 200 MHz), integrating to 10 water protons.

Incremental addition of water, measuring the H_2O integral to assess stoichiometry caused the water resonance to sharpen and shift to higher frequency, as before, and the position of the axial protons shifted towards a limiting value of +98 ppm.

Plotting the change in shift as a function of added water concentration gave a binding isotherm which suggested that the apparent dissociation constant for water exchange in MeCN is 50 ($\pm$ 10) mM.

These observations are compatible with the apparent water dissociation constant of 40 ($\pm$ 10) mM (310 K) determined independently by luminescence studies [39] as discussed in more detail below.

The total concentration of competing ligands, such as water or DMSO, affects the system through two distinct mechanisms. In the first place it perturbs the equilibrium involving the differently solvated species, each of which is characterised by different $\mathcal{D}$ values. Secondly, it affects the rate of solvent exchange between these species. At 300MHz and in the range of temperature observed, the rates are close to the intermediate exchange regime, easily switching from fast to slow exchange.

NIR-CD data

The relationship suggested between the crystal field splitting and the CD pattern deserves further consideration, particularly in relation to the variation of the $\mathcal{D}$ factor revealed by NMR.

The CD spectra of the complexes do not change appreciably on varying the solvent from (wet) acetonitrile to water or (wet) methanol, consistent with the minor changes seen by ^{1}H-NMR. At the same time, the large variations recorded by both spectroscopies upon switching to DMSO solutions (figure 3.35) clearly demonstrate the correspondence between electronic spectra and magnetic properties of the Yb centre. NOE Evidence led us to discard a structural rearrangement as the origin of the change in the magnetic anisotropy constant $\mathcal{D}$, which instead is attributed to a different axial coordination equilibrium in the different solvents. The same mechanism apparently plays a role in determining the energy of the electronic states observed by NIR CD.

It is noteworthy that the spectral width and the amplitude of the signals are comparable in the two solvents (CD_3OD and DMSO) for each species, and that the relative intensities associated with complexes Yb·(R)-**3** and Yb·(S)-**4** are the same in each case. The sign sequence is indifferent to the solvent as well.

Figure 3.36 shows the variation of the CD profile of Yb·(S)-**4** in acetonitrile (10mM) following incremental addition of DMSO. After addition of $10\mu l$ of DMSO, the spectrum became the same as that recorded in pure DMSO solution and did not change any further thereafter. Such behaviour is fully compatible with the NMR data discussed above and supports the hypothesis of a solvent exchange process involving the displacement of the bound H_2O or MeCN ligand by a DMSO molecule.

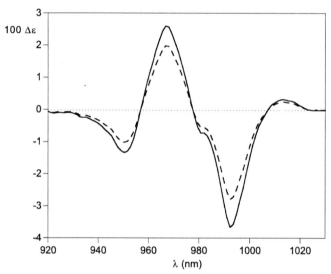

Figure 3.35: NIR-CD spectra of complexes Yb·(R)-**3** (continuous line) and Yb·(S)-**4** (dashed line) in DMSO solutions (c=59 mM and 83 mM; l=1). To make the comparison easier, the curve for Yb·(S)-**4** has been inverted.

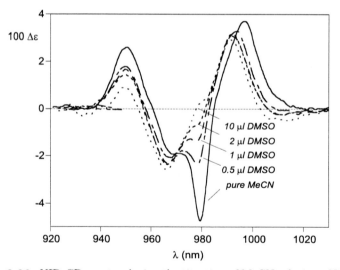

Figure 3.36: NIR-CD spectra during the titration of MeCN solution of Yb·(S)-**4** with DMSO (c=20mM, l=1; DMSO added up from 0.5 μl to 10 μl in a 0.5 ml solution of MeCN).

Luminescence data

Dried samples of Yb·(R)-**3** and Yb·(R)-**4** were dissolved in anhydrous acetonitrile or D_2O (H_2O) to give 1 mM solutions.

The rate of decay of the Yb luminescence was monitored at 310 K following excitation into the phenyl rings at 266 nm [79].

For Yb·(R)-**3**, the rate of decay in H_2O was $1.4 \cdot 10^6$ s^{-1} and fell to $0.13 \cdot 10^6$ s^{-1} in D_2O; with Yb·(R)-**4**, the corresponding figures were $1.25 \cdot 10^6$ s^{-1} and $0.20 \cdot 10^6$ s^{-1} respectively. Given the established sensitivity of the Yb excited state to quenching by OH oscillators, this data is consistent with hydration states of 1.07 and 0.85 for Yb·(R)-**3** and Yb·(R)-**4**, after allowing for the effect of unbound (closely diffusing) OH oscillators which give rise to a quenching effect estimated to be $0.2 \cdot 10^6$ s^{-1} [79].

In dry acetonitrile the rate of decay was measured to be $0.4 \cdot 10^6$ s^{-1} for Yb·(R)-**3** and the same value was recorded for Yb·(R)-**4**. Thus, under these experimental conditions, the data in acetonitrile show that the dried sample is not bound to a water molecule.

Accordingly, water was added incrementally to this solution and the rate of decay of the Yb excited state was observed to increase towards a limit of ca. $1.35 \cdot 10^6$ s^{-1}. By plotting the change in k_{obs} as a function of water concentration and assuming a 1:1 binding stoichiometry, the equilibrium constant for water association (strictly this is for exchange of MeCN by water) was estimated to be 40M^{-1}.

Thus K_d is 25 mM for dissociation of water from Yb·(R)-**3**·H_2O, so that water is not associated significantly at low complex concentrations in dried MeCN. A similar value for K_d of 42 mM has been measured for Eu·(S)-**3** in MeCN [39] by observing changes in the decay of the Eu excited state.

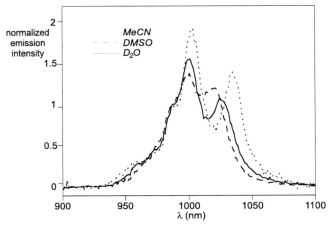

Figure 3.37: Emission spectra for Yb·(R)-**3** recorded in MeCN, D_2O and dry DMSO (310K, 1mM complex, $\lambda_{exc} = 266$ nm).

These changes in axial solvation were also characterised by changes in the form of the emission spectrum of each complex.

The emission spectra for Yb·(R)-**4** (10mM) were measured in dry MeCN, D$_2$O and dry DMSO (figure 3.37), following excitation of the naphthyl chromophore at 305 nm. Very similar spectra were recorded for Yb·(R)-**3**, under identical conditions, following excitation at 255 nm. With the much lower concentrations used to obtain these spectra, there is little chance of adventitious water preferentially associating, and the solvents used were dried to less than 10 ppm water. The emission bands at 980 and 1000 nm varied only slightly in relative intensity. However, the longest wavelength emission band shifted from 1020 nm in DMSO to 1025 in D$_2$O and 1035 nm in MeCN.

The observed solvent dependence is consistent with the behaviour revealed by the near-IR CD studies, and supports the conclusion that the nature of the axial donor is very important in determining the crystal field splitting.

ESI-MS data

The ESI-MS spectrum of the naphthyl derivative Yb·(S)-**4** in water solution is displayed in figure 3.38. The soft ionization employed ensures a minimal perturbation to the system under examination.

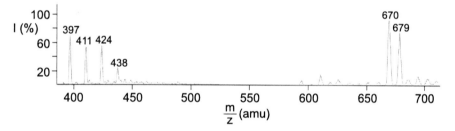

Figure 3.38: ESI-MS spectrum of Yb·(S)-**4** in water solution.

Operating in the low-voltage regime, three sets of peaks can be detected, one for each mono-, bi- and tri-positively charged species.

While m/z 1488 ([Yb·(S)-**4**+2TfO]$^+$) can barely be detected, the signals at m/z 652 and around m/z 400, originated by the di- and tri-charged species [Yb·(S)-**4**+TfO]$^{++}$ and [Yb·(S)-**4**]$^{+++}$, are quite large. This point can be explained considering that in water the triflate counterion is not much associated or it is bound very weakly, and the interaction is broken under the CID conditions. It is noteworthy that the tricharged species (m/z 396) features three additional peaks at m/z 410, m/z 424 and m/z 438. This effect can be justified taking into account the presence of one to three crystallization acetonitrile molecules, which, once the counterion has been removed, stay bound and partially *neutralize* the lanthanide charge.

At low voltage, a second small peak is associated to each of the first two signals: to m/z 670 corresponds m/z 679; aside m/z 410 and m/z 424 we can find (very weak) m/z 416 and m/z 430; these signals are due to the coordination of a water molecule. The triflate in the divalent ion probably protected the weakly coordinating water molecule from CID.

In order to check whether these weak signals are truly connected with a specific axial coordination and to rule out the possibility of non-specific partial desolvation, the behaviour of the system was investigated varying the solvent composition.

It must anyway be noted that, increasing the orifice potential (OR), a progressive reduction of the signal of the complex with the solvent; moreover, the "crystallization adducts" with peaks at m/z 410 and m/z 424 go down, leaving as the only tri-charged species the bare complex at m/z 397; correspondingly, the di-charged species appears at m/z 595 while the adduct with the counterion at m/z 670 gets reduced. At high OR voltages (>85 V), the species at m/z 595 is the only one present in the spectrum.

The spectrum in acetonitrile (figure 3.39 (a)) features a tri-charged species with reduced relative intensity, showing that in this solvent the counter ion is more tightly bound to the positive complex. Furthermore, the absence of a peak at m/z 691 ([Yb·(S)-4 +TfO]$^{++}$+21) indicates that no significant binding to the solvent takes place. This means that either Yb·(S)-4 is eight coordinated in such a solvent or the binding is weaker than that with water and is prone to facile dissociation by the CID. Nevertheless, a small signal of coordinated water can be detected: after addition of water till mM concentrations, it progressively grows up, reaching the same intensity of that in the water solution. This behaviour definitely proves that the interaction detected by ESI-MS is the true, specific, axial binding quantified by luminescence decays and it is not an adduct or an artifact caused by incomplete desolvation.

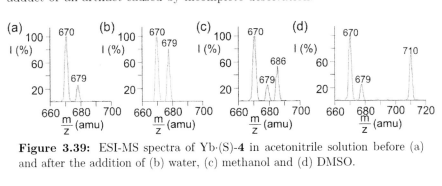

Figure 3.39: ESI-MS spectra of Yb·(S)-4 in acetonitrile solution before (a) and after the addition of (b) water, (c) methanol and (d) DMSO.

On the basis of such findings, we can now investigate the interaction with other solvent molecules, whose effect could only be hypotized and not directly sensed by the other spectroscopic techniques.

Upon adding MeOH to the initial wet MeCN sample (figure 3.39 (b)), a peak at m/z 686 ([Yb·(S)-4+TfO]$^{++}$+16) progressively appears: MeOH coordinates to the lanthanide as well, competing with water for the axial site. It is noteworthy that the orifice voltage necessary for breaking such interaction (about 45 V) is the same required for breaking the water binding, proving that the strenght of the two bonds is comparable. This results is perfectly in-keeping with what observed in the NMR and electronic spectra of the complex (absorbance and CD), which are insensitive to changes from water to methanolic solution).

On the other hand, upon addition of DMSO (figure 3.39 (c)), a peak at m/z 710 ([Yb·(S)-**4**+TfO]$^{++}$+40) progressively grows up and its intensity increases relative to the one of coordinated water at m/z 679 ([Yb·(S)-**4**+TfO]$^{++}$+9): DMSO coordinates to Yb·(S)-**4** and its binding is stronger than water. Two more pieces of information can confirm the latter finding. Firstly, the new peak at m/z 710 disappears at higher orifice voltages (60 eV) than the corresponding limit for the water peak. Moreover, unlike what happened for the other solvents, coordinated species can be revealed also in the tri-charged set of signals, where the m/z 410 and m/z 424 peaks originates two new lines at m/z 430 and m/z 450, which do not dissociate in the conditions of the CID.

3.6.2   YbTHP

An interesting test to check the nature and extent of a possible axial dynamic is a variable solvent study. According to what proposed in Bleaney's model, the magnetic anisotropic susceptibility of the paramagnetic centre, $\mathcal{D}$, is directly dependent on the crystal field experienced by the lanthanide ion [64], which is in turn strongly affected by a ninth capping ligand [72].

Such an effect has been recently demonstrated in the case of two Yb amide complexes, for which the $\mathcal{D}$ parameter, and consequently the spectral width, got reduced after changing the solvent composition from dry acetonitrile to water or DMSO [72].

Anyway, the spectral pattern proved almost identical when solid dried Yb-(S)-THP was dissolved either in dry MeCN or in DMSO. Such finding suggests that the axial coordination equilibrium (if it takes place at all) is shifted toward the free complex and no solvent molecule is sensibly bound to the ytterbium ion in solution; the deprotonation process which occurs at higher pHs is therefore due to a bound hydroxyl group.

The ^{1}H-NMR spectrum of the cerium analogue can afford an interesting comparison. Cerium is the paramagnetic *alter ego* of ytterbium at the beginning of the lanthanide transition: while ytterbium represents a limiting case of paramagnetic lanthanide and its properties are archetypical of the late rare-earths, cerium behaviour in solution, on the contrary, constitutes a second limiting case and is a model for the early lanthanides as far as the coordination properties are concerned.

The ^{1}H-NMR spectrum of Ce-(S)-THP was recorded in MeOD at 300 MHz and is displayed in figure 3.40. Eight resonances can be detected, compatible with a single, paramagnetic, species in solution; the assignments were performed by means of two dimensional correlation spectroscopy (COSY) and NOESY and are reported in figure 3.40.

Unlike Yb^{3+}, Ce^{3+} induced shift is considerably smaller and the spectrum

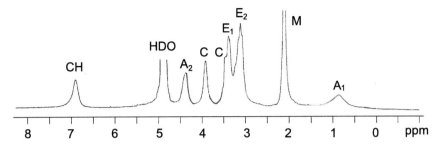

Figure 3.40: ^{1}H-NMR of Ce-(S)-THP in D_2O at pD=4; the peaks are labelled according to figure 3.15.

does not go far beyond the diamagnetic region ranging from 0 to 10 ppm. Moreover, even if the pseudocontact term is still the leading contribution to the whole spectrum, the contact component is no more negligible owing to the globally reduced values of the shifts. So, even though the NMR data of the corresponding lanthanum complex reported in the literature [22] can provide a good description of the diamagnetic reference state of the molecule, no reliable structural optimisation can be performed on the cerium moiety. Some interesting considerations can anyway be drawn on the magnitude of the magnetic anisotropic susceptibility.

Calculations [64] and experimental [80] findings on series of homotypical lanthanide complexes show that the pseudocontact contribution brought about by cerium is about one third of that of Ytterbium provided the geometry of the molecule stays the same. On roughly comparing the spectral widths of the two ^{1}H-NMR spectra recorded in the same conditions, on the contrary, one finds that the ratio is almost one tenth, proving that some structural differences may exist between the two species.

To discern whether this changes involve (only) the conformation of the macrocycle or if different axial dynamics are present on the two ions, a few microliters of DMSO were added to the methanolic solution of Ce-(S)-THP: the spectral profile changed dramatically, with all the resonances exhibiting exchange broadening. Such phenomenon could be frozen at -60°, where the large signals decoalesced into two exchanging families of peaks.

Even if the small and crowded spectral range did not allow any further characterization of such dynamics, it is quite clear that the two species involve different solvent coordination equilibria, with DMSO competing with water for the axial site on the lanthanide.

The experiments described above are thus compatible with a picture in which a solvent molecule is coordinated to the rare-earth ion in Ln-(S)-THP at the beginning of the transition, while the lanthanide is only octa-coordinated toward the end. Indeed, the average trans O-Yb-O bond angle in the calculated structure has a value of 125°, which is sensibly smaller than the value of 137° found in the crystal structure of the Eu complex, and does not allow any further coordination [22].

The model presented here correlates well to the activity of the complexes in promoting the phosphate diester transesterification, which goes in the order $La^{3+} > Eu^{3+} > Lu^{3+}$, in agreement with the reduction of the coordination number [47]. Therefore, the mechanism first hypotized by Chin and Morrow can be confirmed and our analysis proves to be an indirect evidence of an axial binding of the substrate (e.g.: the phosphate ester) during catalysis.

3.7   Generalization to the whole lanthanide transition

Crystal structures of the amide complexes

As anticipated in section 3.2.2, single crystal X-ray diffraction data are available from the literature for the $[Ln \cdot (R)\text{-}3(H_2O)] \cdot CF_3SO_3 \cdot 3H_2O$ complexes (Ln = Eu, Dy and Yb). The molecules crystallized in the chiral space group $P2_12_12_1$ and feature an isostructural nine-coordinated lanthanide. The ligand distances Yb-N and Yb-O get progressively shorter on going from Eu to Yb (Table 3.14), according to the well-known contraction and to reported values for homotypical lanthanide series; on the other hand, $Ln\text{-}OH_2$ distances remain constant within 0.01 Å.

	Eu $\cdot$(R)-**3**	Dy$\cdot$(R)-**3**	Yb $\cdot$(R)-**3**
$\langle Ln\text{-}N \rangle$	2.70	2.65	2.62
$\langle Ln\text{-}O \rangle$	2.37	2.34	2.28
$Ln\text{-}OH_2$	2.43	2.42	2.44

Table 3.14: Relevant average bond lengths (Å) in the Eu $\cdot$ (R)-**3**, Dy $\cdot$ (R)-**3** and Yb $\cdot$ (R)-**3** complexes.

The average N-C-C-N and N-C-C-O torsional angles are +59 ±1 and -29 ±1 for all the three molecules, consistent with the $\Lambda(\delta\delta\delta\delta)$ configuration.

^{1}H-NMR data on the amide complexes

Analysis of the paramagnetic induced shift in different lanthanide complexes according to the procedures described in section 2.1.4 can afford interesting information about possible structural variations occurring in the molecules along the series.

The shifts of the inner portion (the DOTA-part) of Ln·(R)-**4** in methanol were recorded and analized for Ln=Pr, Eu, Dy and Yb; completely analogous results were achieved for the values of the Ln·(R)-**3** series.

The Pr, Eu and Yb complexes exhibit sufficiently short electronic relaxation times to allow the assignment of all the NMR signals using well-resolved two-dimensional COSY and NOESY spectra; the large magnetic moment of Dy, associated with the huge spectral width induced by it, prevents the detection of short range NOE effects and the ^{1}H-NMR spectrum was assigned according to a well-established iterative process [81, 82], described in section 2.1.4, in which the F_i and G_i terms extracted from the analysis of the rest of the series are then used to calculate a predicted spectrum which is compared with the experimental one.

The values of the shift for each assigned signal are summarised in table 3.15.

	Pr·(R)-4	Eu ·(R)-4	Dy ·(R)-4	Yb ·(R)-4	(R)-4
axial 1	-36.3	27.5	-386.6	102.0	1.5
axial 2	11.8	-9.9	107.1	-36.0	1.5
equatorial 1	-1.5	-10.4	-126.5	13.3	1.5
equatorial 2	-4.4	-5.0	-130.7	16.7	1.5
acetate 1	12.6	-16.4	55.9	-31.2	2.4
acetate 2	23.3	-16.8	244.2	-68.9	2.4

Table 3.15: ^{1}H-NMR shifts in d$_4$-methanol for the DOTA-portion of Pr·(R)-4, Eu ·(R)-4, Dy ·(R)-4 and Yb ·(R)-4 complexes, as well as the free ligand (R)-4. The proton labelling is analogous to that used in section 3.4.2.

The analysis of the isotropic paramagnetic shifts according to equation 2.38 leads to the following conclusions:

- the series is isostructural in solution as it appears in the solid state and no significant geometrical changes occur along the transition; the correlation $\frac{\Delta_{ij}}{\langle S_z \rangle_j}$ vs $\frac{\Delta_{kj}}{\langle S_z \rangle_j}$ is very good for all the proton pairs (see, for example, figure 3.41) and can be extended from Pr^{3+} to Yb^{3+}.

proton	F_k (±0.5 %)	R_{ik} (exp) (±0.5 %)	R_{ik} (X-rays) (±0.5 %)
A1	0.36	–	1
A2	-0.48	-0.34	-0.31
E1	-1.69	0.18	0.17
E2	-1.30	0.21	0.22
C1	-1.39	-0.27	-0.29
C2	-0.38	0.67	-0.72

Table 3.16: F_i factors and geometric ratios R_{ik} (i=A1) obtained from plots of $\frac{\Delta_{ij}}{\langle S_z \rangle_j}$ vs $\frac{\Delta_{kj}}{\langle S_z \rangle_j}$ according to equation 2.38 for Ln·(R)-4. The set of geometric ratios R_{ik} of the third column was calculated averaging to C_4 symmetry the crystal data of the Eu^{3+}, Dy^{3+} and Yb^{3+} complexes.

Comparing the fittings from different proton pairs, the F_i and F_k values can be obtained and are reported in Table 3.16. Such table also collects

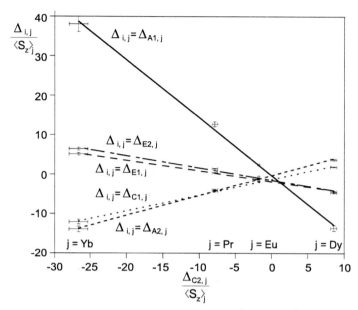

Figure 3.41: Example of correlation between $\frac{\Delta_{ij}}{\langle S_z \rangle_j}$ and $\frac{\Delta_{kj}}{\langle S_z \rangle_j}$ according to equation 2.38 in the four Ln·**4** complexes; Δ_{kj} are here the shifts of the C_2 proton in the different molecules with the lanthanides j, Δ_{ij} are the shifts of another proton i in the same complex with j.

the geometrical factors R_{ik} (defined by equation 2.39) and their expected valus calculated from the C_4- and Ln–averaged X-ray crystal structures of Ln·(R)-**3** (Ln = Eu, Dy, Yb): the calculated and experimental R_{ik} values are in very good agreement.

Unfortunately, because of the perfect equivalence of the shift contributions brought about by the two structural n- and p- types (see point below), the presence of one of the two forms must anyway be checked through NOE spectroscopy case by case. This *Overhauser analysis* could be performed on the Pr^{3+}, Eu^{3+} and Yb^{3+} derivatives, but not on the Dy^{3+} complexes, for the reasons commented at the beginning of the section. Anyway, the occurrence of the n- form before and after Dy^{3+} in the transition can be "interpolated" and suggests that the square antiprism is the conformation of the Dy complex as well.

- the evaluation of the contact contribution for Yb^{3+} allows us to obtain *correct* pseudocontact data for a structural optimization (Table 3.17) and to estimate the relevance of the approximation of $\delta_{con} \simeq 0$ made in section 3.4.

Examining table 3.17, it must be noted that:

   1. the δ_{con} values are not small, and they exceed 4 ppm for the protons

proton	δ_{para}	δ_c	δ_{pc}
A1	100.5	0.9	99.6
A2	-36.0	-1.2	-34.8
E1	13.3	-4.4	17.7
E2	16.7	-3.4	20.1
C1	-31.2	-3.6	-27.6
C2	-68.9	-1.0	-67.9

Table 3.17: Separation of contact and pseudocontact contributions for the paramagnetic shifts of Yb·(R)-**4** in MeCN.

which are *antiperiplanar* with respect to the ion in the correspondent N–C bond; this evidence is therefore a confirmation of the validity of the Karplus relation for the contact terms;

2. anyway, if a fitting is produced on the geometrical parameters of the two *n*- and *p*- forms, the agreement factors R are much better with respect to the case of paragraph 3.4.2, but no additional information can be achieved either on the occurrence of one of the two forms (same agreement for the two) or about small rotations of the side arm (the best agreement is again obtained for a dihedral angle Yb-N-C-C of about 167°).

- it must be noted that the plots of $\frac{\Delta_{ij}}{\langle S_z \rangle_j}$ vs $\frac{C_j}{\langle S_z \rangle_j}$ according to equation 2.35 does not lead to linear results. This latter point seems to be an evidence for variations of the crystal field parameter $A_0^2 \langle r^2 \rangle$ along the transition, as already noted in the cases of another C_4-symmetrical macrocyclic complexes [83] and of C_3-symmetrical nine-coordinated triple helical supramolecules [82, 84].

UV-CD data on the amide series

Figure 3.42 shows the absorbance and CD spectra of Pr·(R)-**4**, Eu·(R)-**4** and Yb·(R)-**4** complexes in acetonitrile solutions between 200 and 450 nm. Analogue results were obtained in methanol solutions. Complexes of ligand **4** feature absorption bands in the UV-Vis region, in correspondence to the naphthalene chromophore transitions [85].

Owing to the dissymmetric disposition of the naphthyl groups in the structure, a rotational strenght is produced. While some weaks CD band are already present in the spectrum of the free ligand (R)-**4** ($|\Delta\epsilon < 10^2|$), in the case of its lanthanide complexes we can find the presence of two well-shaped signals (figure 3.42), with a structure similar to that of a *positive couplet* with crossover at 225 nm, even if the negative band is much more pronounced than the corresponding positive one.

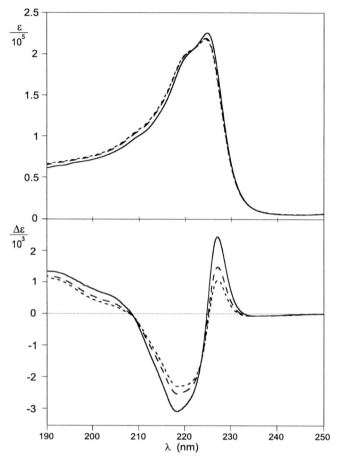

Figure 3.42: UV absorbance (a) and CD (b) of Pr·(R)-**4** (continuous line), Eu·(R)-**4** (dashed line) and Yb ·(R)-**4** (dotted line) in acetonitrile at 25° C (c=10 μM, l=0.01 cm).

First of all, one must note that the spectra have comparable structure and amplitudes, witnessing a strong structural homology in the side arm portion of the molecules as well. The qualitative comparison of the CD spectra in the UV region associated with the naphthyl chromophore therefore provides another confirmation of the isostructurality of the set of complexes.

As far as the intensities of the signals are concerned, we can onserve a progressive decrease of the dissymmetry factor along the transition. It turns out to be quite difficult, however, to translate this phenomenon into a precise structural information. The asymmetric structure of the CD spectrum indicates that an exciton coupling mechanism between the B_b electronic transitions of the naphthalenes cannot quantitatively account for the whole appearence of the spectrum and a simulation of the bands based on a coupled obsciilator model (as that developed by DeVoe[86–88]) is therefore insufficient to reproduce the experimental results.

3.8 Conclusions

We have shown that stereochemical analysis on this class of molecules requires care and the complementary use of different techniques is strongly recommended.

Conclusions on ^{1}H-NMR

1. Pseudocontact shifted ^{1}H-NMR spectra of YbDOTMA have demonstrated that two forms in slow exchange present in solution are in principle both compatible with the same set of geometrical factors for the protons. Two-dimensional EXSY spectra, on the other side, point out that the two forms correspond to different conformations of the macrocyclic ring, while steady-state NOE spectra indicate which is the geometry of the major isomer, thus finally solving the problem of the structural assignment of this molecule in solution. The two species are in diastereomeric relation, as p- or n-forms, but the relative population of the two is reversed with respect to YbDOTA.

   While three different equilibria could be observed for YbDOTA, in the temperature range examined in this work only one has been detected between the two YbDOTMA species, since the presence of the methyl group in the acetate side arms enhances the conformational omogeneity of the chelate system, pushing the arrangement of the pendant groups and favouring one conformation of the cycle ethylene groups.

   The steric hindering affects the kinetic parameters, too, and quantitative EXSY measurements show that the ring inversion is slowed down with respect to the corresponding process in YbDOTA. Thermodynamic parameters of the equilibrium indicate that the structural rearrangement may be accompanied by a hydration process.

   It is noteworthy that the two species in equilibrium are $\Lambda(\delta\delta\delta\delta)$ and $\Lambda(\lambda\lambda\lambda\lambda)$, thus characterized by the same helicity about the metal cation. One can therefore conclude that, in analogy with other chiral ligands, [20, 24, 33] YbDOTMA shows a transfer of chirality from a local stereogenic element to the overall structure. This is a key point in understanding the stereochemistry and consequently the chiroptical properties of this complexes (see 3.5.1).

2. ^{1}H-NMR shift data on the set of amide complexes (Yb-**3** to Yb-**7**) provides a powerful tool for structural refinement but yields information on isomeric forms which cannot be simply discriminated. Only by NOE

measurements can one assign which is the more stable structure in solution.

In spite of the detailed information obtained by these methods, it may seem surprising that the local stereochemistry around the metal ion is still not defined. Indeed, the fact that the stereogenic centre is remote from the inner sphere, justifies the absence of experimental configurational correlation.

3. In the case of YbTHP, ^{1}H-NMR allowed the first complete description of the solution structure of an hydroxyl complex with a transition ion and the analysis of a non-covalent dimeric lanthanide complex.

4. A parallel ^{1}H-NMR investigation of the complexes of (R)-**4** with various lanthanide ions confirmed the isostructurality of the coordination cages of these species along the transition. Such conclusion is compatible with the UV-CD spectra of three complexes of (R)-**4** with Pr^{3+}, Eu^{3+} and Yb^{3+}, which show that, together with a substantial homology, some slight changes of the naphthyl positions could progressively occur within the whole rare-earth series.

Conclusions on NIR-CD

1. Ytterbium (III) ion can be used as a suitable probe to define the stereochemistry of coordinated molecules as its absorptions fall in a spectral region rather devoid of contributions from organic chromophores.

   In addition NIR-CD spectra of Ytterbium complexes can be recorded on suitable dichrographs with relative ease, in contrast to a somewhat common belief, since the anisotropy factors g of several components of the $^2F_{7/2} \rightarrow {}^2F_{5/2}$ term are large. In the examined complexes, the sign sequence of the dichroic bands can be regarded as a direct consequence of the Λ- or Δ-configuration of the coordination polyhedron.

2. The spectrum of YbDOTMA showed several well-resolved transitions around 980 nm with very high dissymmetry factors. On lowering the temperature, the presence of hot bands is highlighted, which leads to a tentative assignement of the transitions.

   The analysis of the spectrum presented here will hopefully serve as a building block to a more non-empirical approach to the problem of the NIR transitions of this ion.

3. In the case of the amide complexes, NIR-CD spectroscopy provides the information which is missing in the NMR context, by relating the spectrum to that of YbDOTMA, which has been fully assigned and confirms the absolute configuration determined by the earlier X-ray crystallographic study.

4. In the case of YbTHP, NIR-CD was used to two different purposes. First of all, the electronic ff transitions enabled us to get information on the equilibrium dynamics of the complex, that is to investigate the number of species in solution and the thermodynamics of their interconversion at different pHs, which could not be studied by other spectroscopic means due to severe broadening of the lines in the NMR spectrum and to the lack of other chromophoric units in the systems. Secondly, the analysis of the CD pattern was employed for a structural determination, being related to the helicity of the coordination polyhedron which wraps around the metal ion. In such occurrence, a convergence was obtained with the NMR study of the solution structure at low pH, which led to a detailed description of the geometry.

Conclusions on the axial dynamics

1. From the data collected on the amide complexes Yb-**3** and Yb-**4**, it is clear that a strong correspondence can be observed between magnetic and optical parameters used in describing the behaviour of lanthanide ions in solution.

Although such a correspondence cannot be fully rationalised, it provides a confirmation of Bleaney's hypothesis [64] which has been tested by thorough experimental studies only in the very recent times [80, 89, 90].

Furthermore, analysing the ^{1}H-NMR spectra of the whole family of DOTA-derivatives, it is evident that there are different processes contributing to the values of the magnetic anisotropy $\mathcal{D}$: structural rearrangements and hydration-dehydration processes (or coordination number). The observed $\mathcal{D}$ factor can be considered as the sum of two terms

$$\mathcal{D} = \mathcal{D}_{lig} + \mathcal{D}_{ax}$$

in which $\mathcal{D}_{ax}$ is a function of the nature of the axial ninth ligand and $\mathcal{D}_{lig}$ is the contribution brought about by the coordination of the macrocyclic ligand. The $\mathcal{D}_{lig}$ term is also sensitive to the shape of the coordination polyhedron, assuming different values in the case of distorted anti-prismatic and regular anti-prismatic coordination geometries.

The simultaneous occurrence of structural rearrangement and hydration/dehydration often reported in the literature has somewhat obscured the latter process in favour of the former. Thus, the variation in the observed $\mathcal{D}$ has been related to a conformational change. Moreover, the similarity between complexes based on N-alkylated tetraazacyclododecanes (cyclen) has perhaps been over-emphasised, by extrapolating data involving complexes with different donor atoms.

We have demonstrated that the parameter $\mathcal{D}$ for Yb(III) is very sensitive to the nature of axial ligand and that no simple structural information can be directly extracted from it.

Although the present study is limited to Yb(III) complexes, the conclusions go beyond the behaviour of this ion: Ytterbium among the lanthanides can be considered an extreme system, since the induced shifts are almost completely dipolar in nature; the results obtained can thus be generalized to all the paramagnetic lanthanides for which the pseudocontact term is not negligible with respect to the contact contribution to the chemical shift, this means all elements besides Pm^{3+}, Sm^{3+}, Eu^{3+} and Gd^{3+}.

2. Examination of luminescence data (i.e. the rate of decay of the Yb emission and the form of the emission spectrum itself) completes the picture about the nature of the axial donor providing a link to the chiroptical results and to the solution NMR interpretation.

3. ESI-MS confirms itself as a very sensitive technique and a particularly powerful one in the study of the solution dynamics and equilibria of labile metal complexes.

We were able to show evidence of such a weak bindings like the axial coordination of solvent by the amide complexes. We collected evidence of 9 coordination of Yb·(S)-**4** in different solvents and were able to provide a scale of different bond strenghts from MeCN, to water and methanol and to DMSO.

We determined that the binding of the triflate counterions to the positively charged complex is solvent dependent and, in its turn, modulates the solvent accessibility to the rare-earth cation. Finally, we collected information on the secondary interactions between the amide side chains and some crystallization molecules, which is an important step toward the understanding of the structure of the second coordination shell of this family of molecules.

A full employment of ESI-MS together with other consolidated techniques (NMR, luminescence, CD) can be encouraged in the study of thermally unstable compounds and in the fields of contrast agents and of metal catalysis.

4. For THP, from the comparison of the different behaviour of the Ce^{3+} and Yb^{3+} species in variable solvent studies, we could prove the lack of axial coordination at the end of the Ln transition, lending an explanation for the absence of catalytic activity of the Yb complex. Furthermore, by demonstrating the origin of the dimerization mechanism of such molecules, we could verify that these complexes are able to generate a net of secondary interactions with even bulky substrates in solution. Such charachteristic opens further interest in this class of compounds, since the large number of interactions undoubtedly favors the specificity in the substrate recognition or the chirality induction in a possible enantioselective catalysis. Thus, provided the metal is accessible enough, these

molecules can prove themselves as potential selective catalysts. To this end, a modification on the side arm (e.q. the substitution of the methyl with a phenyl group) can determine a major or minor opening of the macrocyclic cage, increasing the exposition of the central cation and its capabilities for further coordination.

Even though much work has to be done for the full interpretation of the intimate relationship between electronic and magnetic properties of ytterbium in these and similar compounds, the concerted application of NMR, luminescence and NIR-CD spectroscopy can allow a deeper understanding of the coordination properties of these attractive complexes.

Bibliography

[1] COSTAMAGNA, J.; FERRAUDI, G.; MATSUHIRO, B.; CAMPOS-VALLETTE, M.; CANALES, J.; VILLAGRÁN, M.; VARGAS, J.; AGUIRRE, M. J., *Coord. Chem. Rev.* **2000**, *196*, 125–164.

[2] FALLIS, I. A.; FARRUGIA, L. J.; MACDONALD, N. M.; PEACOCK, R. D., *J. Chem. Soc., Dalton Trans.* **1993**, 2759–2763.

[3] BELAL, A. A.; CHAUDHURI, P.; FALLIS, I.; FARRUGIA, L. J.; HARTUNG, R.; MACDONALD, N. M.; NUBER, B.; PEACOCK, R. D.; WEISS, J., *Inorg. Chem.* **1991**, *30*, 4397–4402.

[4] WHITBREAD, S. L.; POLITIS, S.; STEPHENS, A. K. W.; LUCAS, J. B.; DHILLON, R.; LINCOLN, S. F.; WAINWRIGHT, K. P., *J. Chem. Soc., Dalton Trans.* **1996**, 1379–1384.

[5] DHILLON, R.; STEPHENS, A. K. W.; WHITBREAD, S. L.; LINCOLN, S. F.; WAINWRIGHT, K. P., *J. Chem. Soc., Chem. Commun.* **1995**, 97–98.

[6] DHILLON, R. S.; MADBACK, S. E.; CICCONE, F. G.; BUNTINE, M. A.; LINCOLN, S. F.; WAINWRIGHT, S. P., *J. Am. Chem. Soc.* **1997**, *119*, 6126–6134.

[7] WHITBREAD, S. L.; VALENTE, P.; BUNTINE, M. A.; CLEMENTS, P.; LINCOLN, S. F.; WAINWRIGHT, S. P., *J. Am. Chem. Soc.* **1998**, *120*, 2862–2869.

[8] DHILLON, R.; LINCOLN, S. F.; MADBACK, S.; STEPHENS, A. K. W.; WAINWRIGHT, K. P.; WHITBREAD, S. L., *Inorg. Chem.* **2000**, *39*, 1855–1858.

[9] SMITH, C. B.; WALLWORK, K. S.; WEEKS, J. M.; BUNTINE, M. A.; LINCOLN, S. F.; TAYLOR, M. R.; WAINWRIGHT, K. P., *Inorg. Chem.* **1999**, *38*, 4986–4992.

[10] PITTET, P.-A.; LAURENCE, G. S.; LINCOLN, S. F.; TURONEK, M. L.; WAINWRIGHT, K. P., *J. Chem. Soc., Chem. Commun.* **1991**, 1205–1206.

[11] CARAVAN, P.; ELLISON, J. J.; MCMURRY, T. J.; LAUFFER, R. B., *Chem. Rev.* **1999**, *99*, 2293–2352.

[12] DESREUX, J. F., *Inorg. Chem.* **1980**, *19*, 1319–1324.

[13] SPIRLET, M. R.; REBIZANT, J.; DESREUX, J. F.; LONCIN, M. F., *Inorg. Chem.* **1984**, *23*, 359–363.

[14] AIME, S.; BOTTA, M.; ERMONDI, G., *Inorg. Chem.* **1992**, *31*, 4291–4299.

[15] DUBOST, J.-P.; LEGAR, M.; MEYER, D.; SCHAEFER, M., *C. R. Acad. Sci. Paris, Ser. 2* **1991**, *312*, 349–354.

[16] JACQUES, V.; DESREUX, J. F., *Inorg. Chem.* **1994**, *33*, 4048–4053.

[17] AIME, S.; BARGE, A.; BOTTA, M.; FASANO, M.; AYALA, J. D.; BOMBIERI, G., *Inorg. Chim. Acta* **1993**, *246*, 423–429.

[18] AIME, S.; BOTTA, M.; FASANO, M.; MARQUES, M. P.; GERALDES, C. F. G. C.; PUBANZ, D.; MERBACH, A. E., *Inorg. Chem.* **1997**, *36*, 2059–2068.

[19] AIME, S.; BARGE, A.; BENETOLLO, F.; BOMBIERI, G.; BOTTA, M.; UGGERI, F., *Inorg. Chem.* **1997**, *36*, 4287–4289.

[20] DICKINS, R. S.; HOWARD, J. A. K.; LEHMANN, C. W.; MOLONEY, J.; PARKER, D.; PEACOCK, R. D., *Angew. Chem. Int. Ed. Engl.* **1997**, *36*, 521–523.

[21] MORROW, J. R.; AMIN, S.; LAKE, C. H.; CHURCHILL, M. R., *Inorg. Chem.* **1993**, *32*, 4566–4572.

[22] CHIN, K. O. A.; MORROW, J. R.; LAKE, C. H.; CHURCHILL, M. R., *Inorg. Chem.* **1994**, *33*, 656–664.

[23] AIME, S.; BATSANOV, A. S.; BOTTA, M.; HOWARD, J. A. K.; PARKER, D.; SENANAYAKE, K.; WILLIAMS, J. A. G., *Inorg. Chem.* **1994**, *33*, 4696–4706.

[24] DICKINS, R. S.; HOWARD, J. A. K.; MOLONEY, J. M.; PARKER, D.; PEACOCK, R. D.; SILIGARDI, G., *Chem. Comm.* **1997**, 1747–1748.

[25] HOEFT, S.; ROTH, K., *Chem. Ber.* **1993**, *126*, 869–873.

[26] ELIEL, E. L.; WILEN, S. H.; MANDER, L. N., *Stereochemistry of Organic Compounds;* WILEY: NEW YORK, 1994.

[27] BRITTAIN, H. G.; DESREUX, J. F., *Inorg. Chem.* **1984**, *23*, 4459–4466.

[28] AIME, S.; BARGE, A.; BOTTA, M.; SOUSA, A. S. D.; PARKER, D., *Angew. Chem. Int. Ed. Engl.* **1998**, *37*, 2673–2674.

[29] WOODS, M.; AIME, S.; HOWARD, J. A. K.; MOLONEY, J. M.;

NAVET, M.; PARKER, D.; PORT, M.; ROUSSEAUX, O., *J. Am. Chem. Soc.* **2000**, *122*, 9781–9792.

[30] DI BARI, L.; PINTACUDA, G.; SALVADORI, P., , IN *Book of Abstarcts, 6th International Conference on CD* PISA, ITALY, 1997 .

[31] DI BARI, L.; PINTACUDA, G.; SALVADORI, P., , IN *Book of Abstracts, 24th FGIPS Meeting in Inorganic Chemistry* CORFU, GREECE, 1997 .

[32] DI BARI, L.; PINTACUDA, G.; SALVADORI, P., , IN *Book of abstracts, 23th International Symposium on Macrocyclic Chemistry* TURTLE BAY, HAWAII, 1998 .

[33] HOWARD, J. A. K.; KENWRIGHT, A. M.; MOLONEY, J. M.; PARKER, D.; PORT, M.; NAVET, M.; ROUSSEAU, O.; WOODS, M., *Chem. Comm.* **1998**, 1381–1382.

[34] AIME, S.; BOTTA, M.; FASANO, M.; TERRENO, E.; KINCHESH, P.; CALABI, L.; PALEARI, L., *Magn. Res. Med.* **1996**, *35*, 648–651.

[35] TWEEDLE, M. F.; KUMAR, K.; SHUKLA, R. B.; RANGANATHAN, R.; AIME, S.; BOTTA, M.; MARCO, J. D.; GOUGOUTAS, Z., , IN *Book of abstracts, 23th International Symposium on Macrocyclic Chemistry* TURTLE BAY, HAWAII, 1998 .

[36] DICKINS, R. S.; HOWARD, J. A. K.; MAUPIN, C. L.; MOLONEY, J. M.; PARKER, D.; RIEHL, J. P.; SILIGARDI, G.; WILLIAMS, J. A. G., *Chem. Eur. J.* **1999**, *5*, 1095–1105.

[37] MAUPIN, C. L.; PARKER, D.; WILLIAMS, J. A. G.; RIEHL, J. P., *J. Am. Chem. Soc.* **1998**, *120*, 10563–10564.

[38] AIME, S.; BARGE, A.; BRUCE, J. L.; BOTTA, M.; HOWARD, J. A. K.; MOLONEY, J. M.; PARKER, D.; DE SOUSA, A. S.; WOODS, M., *J. Am. Chem. Soc.* **1999**, *121*, 5762–5771.

[39] BATSANOV, A. S.; BEEBY, A.; BRUCE, J. I.; HOWARD, J. A. K.; KENWRIGHT, A. M.; PARKER, D., *Chem. Comm.* **1999**, 1011–1012.

[40] DI BARI, L.; PINTACUDA, G.; SALVADORI, P., *Eur. J. Inorg. Chem.* **2000**, 75–82.

[41] DI BARI, L.; PINTACUDA, G.; SALVADORI, P., *J. Am. Chem. Soc.* **2000**, *122*, 5557–5565.

[42] AIME, S.; FASANO, M. B. M.; TERRENO, E., *Acc. Chem. Res.* **1999**, *32*, 941–949.

[43] KOBAYASHI, S., *Lanthanides: Chemistry and Use in Organic Synthesis;* SPRINGER-VERLAG: BERLIN, 1999.

[44] BENETOLLO, F.; POLO, A.; BOMBIERI, G.; FONDA, K. K.; VALLARINO, L. M., *Polyhedron* **1990**, *9*, 1411.

[45] LISOWSKI, J.; SESSLER, J. L.; LYNCH, V.; MODY, T. D., *J. Am. Chem. Soc.* **1995**, *117*, 2273–2285.

[46] LISOWSKI, J., *J. Magn. Reson. Chem.* **1999**, *37*, 287–294.

[47] CHIN, K. O. A.; MORROW, J. R., *Inorg. Chem.* **1994**, *33*, 5036–5041.

[48] BELAL, A. A.; FARRUGIA, L. J.; PEACOCK, R. D.; ROBB, J., *J. Chem. Soc., Dalton Trans.* **1989**, 931–935.

[49] FALLIS, I.; FARRUGIA, L. J.; M.McDONALD, N.; PEACOCK, R. D., *Inorg. Chem.* **1991**, *32*, 779–780.

[50] BOLM, C.; KADEREIT, D.; VALACCHI, M., *Synlett* **1997**, 687–688.

[51] MORROW, J. R.; CHIN, K. O. A., *Inorg. Chem.* **1993**, *32*, 3357–3361.

[52] PITTET, P.-A.; FRÜH, D.; TISSIÉRES, V.; BÜNTZLI, J.-C. G., *J. Chem. Soc., Dalton Trans.* **1997**, 895–900.

[53] MORROW, J. R.; AURES, K.; EPSTEIN, D., *J. Chem. Soc., Chem. Commun.* **1995**, 2431–2432.

[54] EVANS, D. A.; NELSON, S. G.; GAGNÉ, M. R.; MUCI, A. R., *J. Am. Chem. Soc.* **1993**, *115*, 9800–9801.

[55] VAN LOOP, A. M.; PETERS, J. A.; VAN BEKKUM, H., , IN *Book of abstracts, 33th International Conference on Coordination Chemistry* FLORENCE, ITALY, 1998 .

[56] DE GRAAUW, C. F.; PETERS, J. A.; VAN BEKKUM, H.; HUSKENS, J., *Synthesis* **1994**, 1007–1017.

[57] EPSTEIN, D. M.; CHAPPELL, L. L.; KHALILI, H.; SUPKOVSKI, R. M.; W. DE W. HORROCKS, J.; MORROW, J. R., *Inorg. Chem.* **2000**, *39*, 2130–2134.

[58] EVANS, C. H., *Biochemistry of the Lanthanides;* PLENUM PRESS: NEW YORK, 1990.

[59] PARKER, D.; WILLIAMS, J. A. G., *J. Chem. Soc., Dalton Trans.* **1996**, 3613–3628.

[60] PARKER, D., *Coord. Chem. Rev.* **2000**, *205*, 109–130.

[61] ALEXANDER, V., *Chem. Rev.* **1995**, *95*, 273–342.

[62] AIME, S.; BOTTA, M.; FASANO, M.; TERRENO, E., *Chem. Soc. Rev.* **1998**, *27*, 19–29.

[63] CHAPPELL, L. L.; D. A. VOSS, J.; W. DE W. HORROCKS, J.; MORROW, J. R., *Inorg. Chem.* **1998**, *37*, 3989–3998.

[64] BLEANEY, B., *J. Magn. Reson.* **1972**, *8*, 91–100.

[65] BERTINI, I.; LUCHINAT, C., *Solution NMR of paramagnetic molecules. Applications to metallobiomolecules and models;* ELSEVIER: AMSTERDAM, 2001.

[66] PERRIN, C. L.; GIPE, R. K., *J. Am. Chem. Soc.* **1984**, *106*, 4036–4038.

[67] ABEL, E. W.; COSTON, T. P. J.; ORREL, K. G.; IK, V.; STEPHEN-

SON, D., *J. Magn. Reson.* **1986**, *70*, 34–53.

[68] AIME, S.; BOTTA, M.; ERMONDI, G.; TERRENO, E.; ANELLI, P. L.; FEDELI, F.; UGGERI, F., *Inorg. Chem.* **1996**, *35*, 2726–2736.

[69] AIME, S.; BOTTA, M.; PARKER, D.; WILLIAMS, J. A. G., *J. Chem. Soc., Dalton Trans.* **1995**, 2259–2266.

[70] HÜFNER, S., *Optical spectra of transparent rare earth compounds;* ACADEMIC PRESS: NEW YORK, 1978.

[71] BODENHAUSEN, J.; ERNST, R.; WOKAUN, D., *Principles of Nuclear Magnetic Resonance in One and Two Dimensions;* CLARENDON PRESS: OXFORD, 1990.

[72] DI BARI, L.; PINTACUDA, G.; SALVADORI, P.; PARKER, D.; DICKINS, R. S., *J. Am. Chem. Soc.* **2000**, *122*, 9257–9264.

[73] SALVADORI, P.; ROSINI, C.; BERTUCCI, C., *J. Am. Chem. Soc.* **1984**, *106*, 2439–2440.

[74] MESSORI, L.; MONANNI, R.; SCOZZAFAVA, A., *Inorg. Chim. Acta* **1986**, L15–L17.

[75] MASON, S., *Molecular optical activity and the chiral discriminations;* CAMBRIDGE UNIVERSITY PRESS: CAMBRIDGE, 1982.

[76] REID, M. F.; RICHARDSON, F. S., *J. Phys. Chem.* **1984**, *88*, 3579–3586.

[77] SPAULDING, L.; BRITTAIN, H. G., *Inorg. Chem.* **1983**, *22*, 3486.

[78] ZHANG, X.; CHANG, C. A.; BRITTAIN, H. G.; GARRISON, J. M.; TELSER, J.; TWEEDLE, M. F., *Inorg. Chem.* **1992**, *31*, 65597.

[79] BEEBY, A.; CLARKSON, I. M.; DICKINS, R. S.; FAULKNER, S.; PARKER, D.; ROYLE, L.; DE SOUSA, A. S.; WILLIAMS, J. A. G., *J. Chem. Soc., Perkin Trans. 2* **1999**, 493–503.

[80] BERTINI, I.; YANIK, M. B. L.; LEE, Y.-M.; LUCHINAT, C.; ROSATO, A., *J. Am. Chem. Soc.* **2001**, *123*, 4181–4188.

[81] RIGAULT, S.; PIGUET, C.; BÜNTZLI, J.-C. G., *J. Chem. Soc., Dalton Trans.* **2000**, 2045–2053.

[82] RIGAULT, S.; PIGUET, C., *J. Am. Chem. Soc.* **2000**, *122*, 9304–9305.

[83] REN, J.; SHERRY, A. D., *J. Magn. Reson. B* **1996**, *111*, 178–182.

[84] PIGUET, C.; EDDER, C.; RIGAULT, S.; BERNARDINELLI, G.; BÜNTZLI, J.-C. G.; HOPFGARTNER, G., *J. Chem. Soc., Dalton Trans.* **2000**, 3999–4006.

[85] STREITWIESER, JR., A., *Molecular orbital theory (for organic chemists);* WILEY: NEW YORK, 1961.

[86] DEVOE, H., *J. Chem. Phys.* **1965**, *43*, 3199.

[87] CECH, C. L.; HUG, W.; TINOCO, JR., I., *Biopolymers* **1976**, *15*, 131.

[88] HUG, W.; CIARDELLI, F.; TINOCO, JR., I., *J. Am. Chem. Soc.* **1974,** *96,* 3407.

[89] BABUSHKINA, T. A.; ZOLIN, V. F.; KORENEVA, L. G., *J. Magn. Reson.* **1983,** *52,* 169–181.

[90] BRUCE, J. I.; PARKER, D.; TOZER, D. J., *Chem. Commun.* **2001,** 2250–2251.

Chapter 4

CD spectra calculations on YbDOTMA

During the seventies and the eighties, Peacock, Mason and Stewart [1] and Reid and Richardson [2, 3] developed a model for the interpretation of absorption transitions and CD bands of trivalent lanthanide ion complexes. Usually the adopted scheme represents the metal ion and the ligands as independent subsystems and considers their interactions as purely electrostatic. In this way, a perturbation originates which makes electrically allowed a first-order forbidden transition. The frequency and the intensity of this transition depend on the geometric and electronic properties of the ligand system.

The goal of such approach is to parametrize the absorption data in a scheme general enough to cover in detail all the ion-ligand-radiation interaction mechanisms. Thus, the phenomenological parameters obtained this way would provide macroscopic structural correlations applicable to wide classes of systems and independent on the physical modalities of the interaction. This theoretical model has only been employed for the study of the electronical properties of families of crystalline systems, chiral only in the solid state and charachterized through diffractometric methods [4–6]. Most of the relevance of lanthanide complexes in chemistry or in biological studies on the contrary is connected to molecules in *solution* and this is the state in which the structural determination is relevant. Unfortunately, most optical activity data on Ln^{3+} systems in solution media have been obtained on complexes where coordination numbers and metal-ligand stoichiometries are either ill-defined or unknown. A first step to build a conection between CD spectrum and structure is therefore the analysis of stable and relatively simple complexes and the class of DOTA-derivatives studied in this thesis provides a basis for this kind of studies, working as *prototypical* model systems.

In the previous chapter, the electronic NIR-CD spectra of the ff transition of Yb^{3+} provided a relevant contribution to the structural determination of the complexes containing the lanthanide ion. In those analyses, CD allowed the description of the twist of the ligand around the metal by means of *empirical* correlations with a reference geometry of known distortion.

Here a different purpose is looked for with the opposite attitude. Starting from a well-determined solution geometry, we will try to calculate, on *semi-empirical* grounds, the shape and the signs of the CD bands.

Such an effort will therefore be a way to take advantage of the model developed by the groups of Mason and Richardson for the expressions of the dipole and rotatory strenghts of ff transitions of lanthanides, examining its sensitivity to various chemical and structural features of the ligand environment about the metal ion.

4.1   Parameters employed

The YbDOTMA (Yb·(R)-**2**) structural parameters used in the present study were deduced by the NMR-optimized solution geometry of its major form described in section 3.4.1. In the model adopted, the chemical structure of the ligand environment about the Yb^{3+} ion and the chemical identities of the individual ligand atoms are determined entirely by ligand charges (q_L) and mean ligand polarizabilities ($\bar{\alpha}_L$) endowed with a "point nature".

Point charges of the free ligand were estimated by a CNDO calculation; the atomic polarizabilities were assigned according to Applequist's "atom dipole interaction" model for molecular polarizabilities [7].

While for many of the carbon and hydrogen atoms of the macrocycle the q_L and $\bar{\alpha}_L$ resulted negligible, the values of Table 4.1 were assigned to the carboxilates and the ring nitrogens. Their position relative to the Yb^{3+} cation is also reported.

Table 4.2 shows the values of the spin-orbit constant, taken from [8], and the radial integrals for the calculation of the crystal field perturbation potential (equation 1.19), taken from Watson and Freeman [9]. Displayed in Table 4.2 are also the $\langle 4f|r^{2i+1}|5d\rangle$, $\langle 4f|r^{2i+1}|n'g\rangle$, $\langle 4f|r^{2i}|4f\rangle$, $\Delta(5d)$ and $\Delta(n'g)$ values necessary for the computation of the static and dynamic electric dipole moments. Since their values were missing in Krupke's data [10], they were extrapolated from the available set of lantanide ions, ranging from Pr^{3+} to Tm^{3+}.

4.2   Free-ion wave functions

In the case of Yb^{3+}, the *free ion* electronic Hamiltonian, constructed in the Russell-Saunders basis $|\psi\left[\frac{1}{2}3\right]\rangle$), is constituted only by the spin-orbit coupling between the angular and spin orbital moments $\mathbf{l}_i$ and $\mathbf{s}_i$ of the the i electrons:

$$\mathcal{H}_{so} = \sum_{i=1}^{N} \zeta_{so}\,\mathbf{s}_i \cdot \mathbf{l}_i \tag{4.1}$$

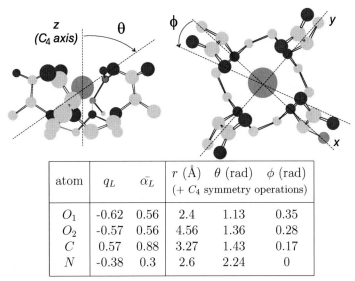

atom	q_L	$\bar{\alpha}_L$	r (Å)	θ (rad)	ϕ (rad)
			($+$ C_4 symmetry operations)		
O_1	-0.62	0.56	2.4	1.13	0.35
O_2	-0.57	0.56	4.56	1.36	0.28
C	0.57	0.88	3.27	1.43	0.17
N	-0.38	0.3	2.6	2.24	0

Table 4.1: Ligand parameter set for Yb·(R)-**2** used in the calculation. O_1 is the carboxylate coordinated oxygen, O_2 the uncoordinated carboxilate oxygen, C the carboxilate carbon and N the ring nitrogen.

parameter	value		ref.
ζ_{so}	10250	cm^{-1}	[8]
$\langle 4f|r^2|4f\rangle$	0.171	Å^2	[9]
$\langle 4f|r^4|4f\rangle$	0.075	Å^4	[9]
$\langle 4f|r^6|4f\rangle$	0.068	Å^6	[9]
$\langle 4f|r^8|4f\rangle$	0.092	Å^8	[10]
$\langle 4f|r|5d\rangle$	0.290	Å	[10]
$\langle 4f|r^3|5d\rangle$	0.318	Å^3	[10]
$\langle 4f|r^5|5d\rangle$	0.542	Å^5	[10]
$\Delta\,(5d)$	$1.00 \cdot 10^5$	cm^{-1}	[10]
$\Delta\,(n'g)$	$2.17 \cdot 10^5$	cm^{-1}	[10]

Table 4.2: Values for Yb^{3+} ion parameters.

By diagonalization of the Hamiltonian 4.1, the resulting wavefunctions, expressed as $|A_m\rangle = |\psi\left[\frac{1}{2}3\right] J\,M_J\rangle$, are constituted by two sets of degenerate levels ($J = \frac{7}{2}$ and $\frac{5}{2}$), featuring eigth ($M_J = 7/2, ..., -7/2$) and six ($M_J = 5/2, ..., -5/2$) components respectively and separated by the radial spin orbit coupling constant ζ_{so}.

4.3 Crystal field parameters and transition frequencies

The "charge and polarizability" model by Richardson [11, 12] was employed to describe the environment surrounding the Yb^{3+} ion.

From the parameters described in the previous paragraph, the set of crystal field parameters of table 4.3 has been derived, using the relations 1.17 and 1.21 of pag. 8.

parameter (cm^{-1})	contributions		
	q	$\bar{\alpha}_L$	total
B_0^2	830	1800	2630
B_0^4	-130	-175	-305
$B_{\pm 4}^4$	$30 \mp 125\,i$	$115 \mp 230\,i$	$145 \mp 355\,i$
B_0^6	-25	-50	-75
$B_{\pm 4}^6$	$-12 \pm 20\,i$	$25 \pm 30\,i$	$-37 \pm 50\,i$

Table 4.3: Calculated crystal field parameters.

The crystal field Hamiltonian calculated in the Russel-Saunders basis according to equation 1.22 is given by:

$$\mathcal{H}_{cf}(\tfrac{7}{2}) = \begin{array}{c|cc|cc} & \pm 7/2 & \mp 1/2 & \pm 5/2 & \mp 3/2 \\ \hline \pm 7/2 & 616.0 & 13.1 + 36.6\,i & & \\ \mp 1/2 & 13.1 - 36.6\,i & -486.5 & & \\ \hline \pm 5/2 & & & 141.2 & 26.2 + 63.0\,i \\ \mp 3/2 & & & 26.2 - 63.0\,i & -270.6 \end{array}$$

for the $^2F_{7/2}$ and by:

$$\mathcal{H}_{cf}(\tfrac{5}{2}) = \begin{array}{c|cc|c} & \pm 5/2 & \mp 3/2 & \pm 1/2 \\ \hline \pm 5/2 & 616.0 & 13.1 + 36.6\,i & \\ \mp 3/2 & 13.1 - 36.6\,i & -486.5 & \\ \hline \pm 1/2 & & & 141.2 \end{array}$$

for the $^2F_{5/2}$ manifold.

By diagonalizing the Hamiltonian, the following crystal field function set is obtained for the $4f$ electrons of Yb^{3+}:

where each function has the shape of:

$$|A_b) = \sum_m C_{bm}|A_m) = \sum_{\psi[SL]JM_J} C_b(\psi[SL]JM_J)|\psi[SL]JM_J), \qquad (4.2)$$

Level	E (cm^{-1})	CF wavefunction	
3′ ————	466	$\|\pm 1/2)$	
2′ ————	77	$0.995\|\pm 3/2) + [0.037 \mp 0.092\,i]\,\|\mp 5/2)]$	} 2F 5/2
1′ ————	-542	$0.995\|\pm 5/2) - [0.037 \pm 0.092\,i]\,\|\mp 3/2)]$	
4 ————	617	$0.999\|\pm 7/2) - [0.012 \pm 0.033\,i]\,\|\mp 1/2)]$	
3 ————	152	$0.987\|\pm 5/2) - [0.061 \pm 0.147\,i]\,\|\mp 3/2)]$	} 2F 7/2
2 ————	-282	$0.987\|\pm 3/2) + [0.061 \mp 0.147\,i]\,\|\mp 5/2)]$	
1 ————	-488	$0.999\|\pm 1/2) + [0.012 \mp 0.033\,i]\,\|\mp 7/2)]$	

and the c_{bm} are the crystal field coupling coefficients.

The corresponding set of transition frequencies (in nm) can be derived:

	1′	2′	3′
4	1096	1027	987
3	1043	980	944
2	998	940	907
1	978	922	890

Table 4.4: Calculated transition frequencies between CF states of the $^2F_{7/2}$ (vertical, levels 1, 2, 3, 4) and CF states of the $^2F_{5/2}$ (horizontal, states 1′, 2′, 3′).

4.4 Dipole and rotatory strengths

As discussed in chapter 1.2, once the magnetic and electric transition dipole moments have been worked out, the dipole strength and the rotatory strength for each $0 \to a$ ff crystal field transition are given by:

$$D_{0a} = \left(|\mathbf{P}_{0a}|^2 + |\mathbf{M}_{0a}|^2\right) \tag{4.3}$$

$$R_{0a} = Im(\mathbf{P}_{0a} \cdot \mathbf{M}_{0a}) \tag{4.4}$$

The calculation of the magnetic transition dipoles between the crystal field states is straightforward, using eq. 1.34 with $J = \frac{7}{2}$, $J' = \frac{5}{2}$ and $([SL]\frac{7}{2}\|\hat{\mathbf{m}}\|[SL]\frac{5}{2}) = -6.01 \cdot 10^{-20}$ erg/gauss according to eq. 1.37.

The set of the electric transition coefficients $\mathcal{A}$ and $\mathcal{B}$ (eq. 1.41 and 1.42) was calculated from the geometry of the complex, using the values of charges and polarizabilities of table 4.2.

In table 4.5, the dipole strengths D and rotational strengths R associated with each of the transitions frequencies of table 4.4 are reported, separately for the static and dynamic mechanisms (superscript s and d, respectively) described in section 1.2.

$$D^{(s)} = \begin{array}{c|ccc} & 1' & 2' & 3' \\ \hline 4 & 1630 & 47 & 26 \\ 3 & 1760 & 53 & 1 \\ 2 & 1055 & 74 & 107 \\ 1 & 1 & 456 & 409 \end{array} \qquad D^{(d)} = \begin{array}{c|ccc} & 1' & 2' & 3' \\ \hline 4 & 0 & 0 & 17 \\ 3 & 474 & 8 & 0 \\ 2 & 61 & 97 & 0 \\ 1 & 0 & 0 & 11 \end{array}$$

$$R^{(s)} = \begin{array}{c|ccc} & 1' & 2' & 3' \\ \hline 4 & -51.1 & 51.0 & 0.1 \\ 3 & 78.5 & -92.3 & 13.8 \\ 2 & -28.1 & 41.9 & -13.8 \\ 1 & 0.6 & -0.5 & -0.1 \end{array} \qquad R^{(d)} = \begin{array}{c|ccc} & 1' & 2' & 3' \\ \hline 4 & 0 & 0 & 10.0 \\ 3 & 57.7 & 38.7 & 0 \\ 2 & -5.9 & -90.6 & 0 \\ 1 & 0 & 0 & -10.1 \end{array}$$

Table 4.5: Calculated dipole (D) and rotatory strenghts (R), in Debye^2, for the twelve crystal field transitions of YbDOTMA; on the left the results coming from the static model, on the right the ones coming from the dynamic mechanism.

The resulting aborption and circular dichroism spectra associated with the whole multiplet can be then simulated

- weighting each transition component for the Boltzmann population of the starting crystal-field sublevel;

- approximating each transition line with a Lorentzian:

$$\epsilon = \frac{\epsilon_i^{max}}{\left(\frac{x-\bar{\nu}_i}{\bar{\nu}_i}\right)^2 + 1} \qquad \Delta\epsilon = \frac{\Delta\epsilon_i^{max}}{\left(\frac{x-\bar{\nu}_i}{\Delta\bar{\nu}_i}\right)^2 + 1}$$

For a given half-heigth width $\frac{\Delta\bar{\nu}}{2}$, the corresponding ϵ^{max} and $\Delta\epsilon^{max}$ can be worked out from equations 1.26 and 1.27, which for a Lorentzian lineshape become:

$$D_{0a} = 9.18 \cdot 10^{-3} \int \frac{\epsilon}{\bar{\nu}} d\bar{\nu} \simeq \frac{9.18 \cdot 10^{-3}}{\bar{\nu}} \int \epsilon \, d\bar{\nu} = \frac{9.18 \cdot 10^{-3}}{\bar{\nu}} \pi \epsilon^{max} \Delta\nu \qquad (4.5)$$

$$R_{0a} = 0.248 \int \frac{\Delta\epsilon}{\bar{\nu}} d\bar{\nu} \simeq \frac{0.248}{\bar{\nu}} \int \Delta\epsilon \, d\bar{\nu} = \frac{0.248}{\bar{\nu}} \pi \Delta\epsilon^{max} \Delta\nu \qquad (4.6)$$

Figure 4.1 was calculated from the data of table 4.5 using a half-height width $\Delta\tilde{\nu}$ of about 30 nm. Here again the different contributions of the static and dynamic mechanisms are evaluated separately. The experimental spectra are also displayed.

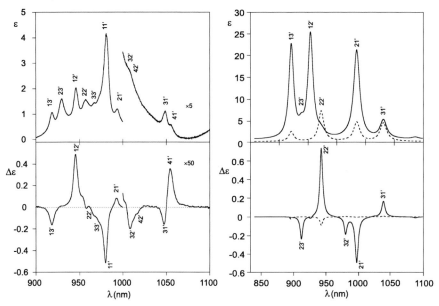

Figure 4.1: Experimental (left) and symulated (right) absorption and CD spectra of YbDOTMA. In the panels on the right, the contributions of the static (continuous line) and dynamic (dashed line) mechanisms are reported separately. The labelling in the panels on the left is the tentative one proposed in section 3.5.1.

4.5 Discussion

A first consideration regards the crystal field coefficients. It is interesting the comparison between the B_0^2 calculated above and the exstimate of the same value given by the susceptibility anisotropy $\mathcal{D}$ which comes from NMR. According to Bleaney [13], the two figures should differ by a factor of about 5; so $\mathcal{D} \simeq 3500$ ppm $\cdot$ Å^3 (see section 3.4.1) should relate to a B_0^2 of about 700 cm^{-1}.

It is clear therefore that the magnitude of the crystal field has been somewhat overestimated. A possible cause of this behaviour, already noted by Richardson [12], can be traced back in the use of Freeman and Watson's $4f$-radial integrals [9] without shielding corrections, which can sensibly alterate the $\langle 4f|r^i|4f\rangle$ values. Indeed, comparing such results with the experimental frequencies, we can note that the calculated set cover a larger spectral window, as a consequence of the stronger predicted crystal field.

As far as the transition dipole moments are regarded, the absorption spectrum does not feature a particular analogy with the experimental counterpart, and the comparison is made even harder by the large crowding of the spectral window in the experimental pattern.

On the contrary, the comparison of the calculated CD with the experimental spectrum is encouraging, since the correspondence between the two is striking. It is noticeable that the static contribution alone can account for the main features of the experimental CD spectrum, the dynamic mechanism being about one order of magnitude smaller. Even though the transitions span a wider range of frequencies, both the amplitude of the signals and their sign alternation are in excellent correspondence with the experimental data.

A disagreement can be noted between the labelling of the calculated transitions and that obtained from the assignem,ent described in section 3.5.1. Such conflict is more apparent rather than substantial and, on the contrary, the mismatched can be smoothed or even completely solved on the basis of the following consideration. In both schemes, all three main CD bands belong to transitions originating from the same sublevel: that is number 2 in the calculated spectrum, while it had assumed to be number 1 in the tentative experimental labelling. The same argument can be used with the second set of three transitions, belonging to sublevel 3 in the calculations and to sublevel 2 in the previous assignement. Both the labelling are thus compatible with the VT experiments and both consistently explain the main features of the recorded CD spectrum, the difference being thus limited to the assignements of minor components of the spectrum.

4.6   Conclusions

The computational model used in this study is faithful to both the formal and the physical aspects of the theoretical model presented by Richardson [11].

The results achieved here suggest that the model employed here is a promising tool which can be used and refined for the quantitative interpretation of the large amount of data which have been collected on lanthanide complexes in solution. The conclusions are even more valuable if we consider that in the past direct calculations of ff electric dipole strengths in chiral lanthanide complexes have achieved, at best, only modest success.

Bibliography

[1] MASON, S. F.; PEACOCK, R. D.; STEWART, B., *Mol. Phys.* **1975**, *30*, 1829–1841.

[2] REID, M. F.; RICHARDSON, F. S., *J. Phys. Chem.* **1984,** *88,* 3579–3586.

[3] RICHARDSON, F. S.; BERRY, M. T.; REID, M. F., *Mol. Phys.* **1986,** *58,* 929–945.

[4] MORAN, D. M.; RICHARDSON, F. S., *Inorg. Chem.* **1992,** *31,* 813–818.

[5] BERRY, M. T.; SCHWIETERS, C.; RICHARDSON, F. S., *Chem. Phys.* **1988,** *122,* 125–139.

[6] MAY, P. S.; REID, M. F.; RICHARDSON, F. S., *Mol. Phys.* **1987,** *62,* 341–364.

[7] APPLEQUIST, J., *Acc. Chem. Res.* **1977,** *10,* 79–85.

[8] CARNALL, W. T.; WYBOURNE, B. G.; FIELDS, P. R., *J. Chem. Phys.* **1965,** *42,* 3797–3805.

[9] FREEMAN, A. J.; WATSON, R. E., *Phys. Rev.* **1962,** *127,* 2058–2075.

[10] KRUPKE, W. F., *Phys. Rev.* **1966,** *145,* 325–337.

[11] RICHARDSON, F. S.; FAULKNER, T. R., *J. Chem. Phys.* **1982,** *76,* 1595–1606.

[12] SAXE, J. D.; FAULKNER, T. R.; RICHARDSON, F. S., *J. Chem. Phys.* **1982,** *76,* 1607–1623.

[13] BLEANEY, B., *J. Magn. Reson.* **1972,** *8,* 91–100.

Chapter 5

Lanthanide complexes as catalysts

During the last decade, the rare earth elements have given an enourmous contribution to the field of organic synthesis, including stereoselective catalysis [1]. The compounds employed in modern organic transformations range from highly efficient inorganic reagents, such as $SmI_2(thf)_2$ or $Ln(OTf)_3$ - commercially available - to more sophisticated organometallic reagents, such as cyclopentadienyl derivatives, prepared on a laboratory scale and often extremely sensitive to air and water. A particularly interesting class of Ln(III) derivatives includes "pseudo-organometallic" compounds, highly reactive metallorganic molecules, which do not contain any direct metal carbon linkage, but only readily hydrolyzable Ln-X bonds. Such compounds (e.g.: amide and alcoxide derivatives) can be important synthetic precursors and feature excellent catalytic behaviour in organic reactions as well [2, 3].

5.1 Heterobimetallic Ln M_3BINOL_3 complexes

Chiral heterobimetallic lanthanide-alkali metal complexes of formula LnM_3-$BINOL_3$ (Ln: lanthanide (III); M: Li, Na, K; BINOL: 1,1'-bi-(2-naphthol)) developed by Shibasaki *et al.*[1, pag.199–232] are endowed with a multifunctional ability to catalyze different types of enantioselective reactions. Indeed, they have recently been employed as catalysts in a large variety of asymmetric syntheses [1, 4, 5], including many "classical" C–C bond formations like the nitroaldol reaction, the Michael addition, the direct aldol reaction and so on, but also an oxidation reaction and the asymmetric formation of C–P bonds (see figure 5.1).

Such new and innovative "broad-band" kind of catalysts contain a Lewis acid as well as a Brønsted base moiety and show a similar mechanistic effect as observed in enzyme chemistry. The catalytic cycle proposed for the direct

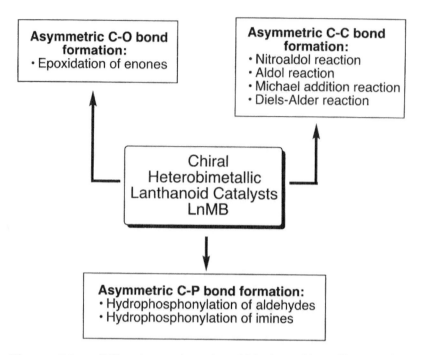

Figure 5.1: Different reactions in which heterobimetallic complexes LnM₃BINOL₃ have proved efficient catalysts.

catalytic asymmetric aldol reaction promoted by LaLi₃BINOL₃ (figure 5.2) highlights the different roles played by the Ln(III) centre - as a Lewis acid activating the aldehyde - and by the Li–binaphthoxide moiety - as a Brønsted base -, allowing the reaction to proceed without the activation of the starting material.

The orientation of the carbonyl function is controlled by the arrangement of the environment of the Ln(III) ion. All of these reaction proceed with high stereoselectivity, resulting in the formation of the desired non racemic products in high to excellent enantiomeric purity. These complexes can be easily synthetized from the corresponding lanthanide trichlorides and/or lanthanide isopropoxides [4, 6, 7].

Investigations concerning the influence of the metal components have been extensively carried out [4]; such analyses showed pronounced diffrences in the reactivity and in the enantioselectivity, both among the various rare earth metals and among the alkali metal used. A great variety of conditions was observed in the different reactions: in the nitroaldol [6], for example, Li proved the best effective alkali partner for Ln(III) and small changes in the ionic radius of the rare earth cation caused drastic changes in the enantiopurity of the produced nitroaldols, with low e.e. at the end of the transition. The trend

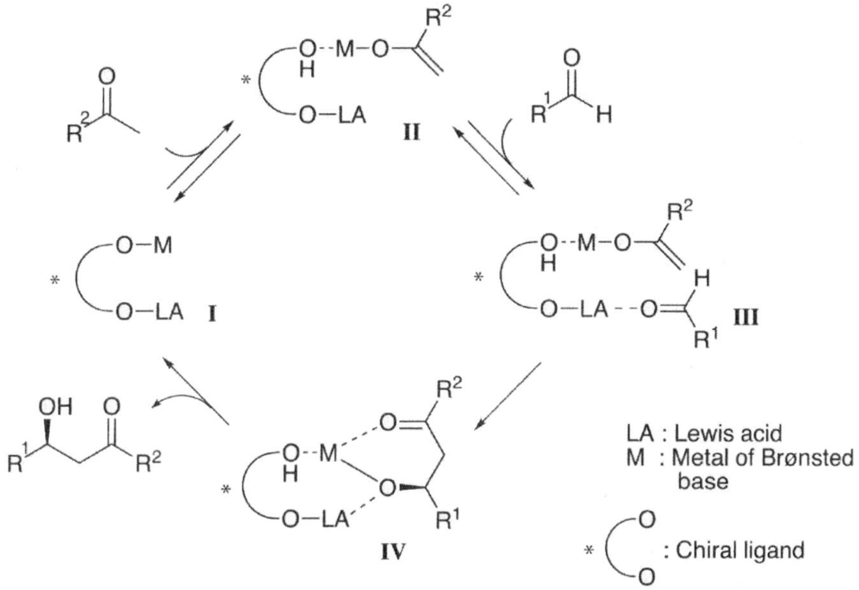

Figure 5.2: Proposed catalytic cycle for the direct catalytic asymmetric aldol reaction (from ref. [1, pag.214]).

was almost reversed in the case of the hydrophosphomylation of cyclic imines [8], where maximum e.e. were obtained with the YbK_3BINOL_3 system.

These heterobimetallic complexes are stable in organic solvents such as THF, CH_2Cl_2 and toluene, which contain small amounts of water and can also stand oxygen; it is noteworthy that they represent one of the few examples in which "real catalitically active species" can be isolated, crystallized or analized by NMR spectroscopy and by MS spectrometry in solution.

Intense investigations focused on the determination of the structure have therefore been carried out. The complete stoichiometry of these species was determined by a LDI-TOF mass spectral analysis and then confirmed by the crystal structures of the $LnNa_3BINOL_3$ (Ln=La,Pr,Nd,Eu) [4–6] (see the following section 5.2 for a description).

Correspondently, a NMR spectroscopic as well as FAB and ESI mass spectrometry study of the isolated YbK_3BINOL_3 complex provided a clear picture into the assembly of this molecule in solution [8].

Anyway, a proper confirmation that the solution structure of these molecules corresponds to that determined in the solid state has never been given. Such a task can be easily performed by analizing shifts and relaxation data of the ytterbium complex, and combining that study with other two major ssources of information, that is the CD spectroscopy in the UV - which concerns the transitions of the binaphtholate chromophore - and in the NIR - which relies on the Yb(III) ff transition.

To this aim, the Na and K derivatives of YbM$_3$BINOL$_3$ were synthetized. The combined use of the aforementioned techniques, beside giving a clear insight into the structure of these complexes, allows one to point out and critically evaluate the role of the alkali metal partner on the chemical properties of these molecules and, consequently, on the spevtroscopic properties of the lanthanide ion.

The crystal structure of YbNa$_3$BINOL$_3$ was also obtained in the course of this work, and constitutes an useful starting point for the analysis described in the following.

5.2   Crystal structure of Yb Na$_3$BINOL$_3$

The crystal structure of the Yb complex is displayed in figure 5.3 and shows that the stoichiometry is (S)-Yb Na$_3$(BINOL)$_3$· 6 THF.

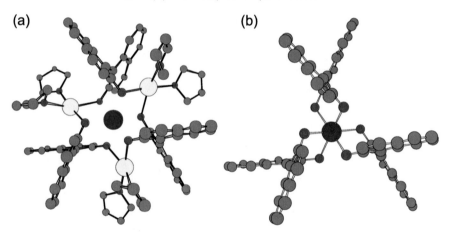

(a)   **(b)**

Figure 5.3: Crystal structure of the Yb Na$_3$BINOL$_3$ complex, with the coordinated THF molecules and the Na$^+$ co-cation displayed (a) and with the lanthanide coordination geometry highlighted (b).

The geometry of the oxygen atoms around the central cation is almost exactly octahedric and the diedral angle θ defined by the two bound nafthalenic unities is 63°. The three sodium ions are coordinated to the oxygens involved in the chelation of the lanthanide ion. Furthermore, six molecules of THF are coordinated to the sodium ions. From this double complexation two different bond distances arise between O and the lanthanide ion, that is 2.203 and 2.230 $\mathring{A}$ in the case of Yb. It is noteworthy that the synthesis of the complex from (S)-BINOL yields only the Λ configuration in the final product, thus showing an interesting case of chirality transfer from the ligand to the whole coordination polyhedron. It will be a matter of interest to check if and how the distortion of the coordination geometry reflects in the electronic structure of the rare earth metal (see part 5.5). The crystal structure obtained is very similar to those reported for the analogue La^{3+}, Pr^{3+}, Nd^{3+} and Eu^{3+} complexes [6], even if in those cases a molecule of water enters axially into the coordination polyhedron. The absence of this further coordination site must not surprise, since the fist part of the f transition easily exhibits higher coordination numbers than the final part; in analogy to other isostructural series [9, 10], it must be noted that the oxygen-to-lanthanide distances shorten up progressively along the family (see table 5.1) till the axial substituent is ejected from the first coordination sphere.

	$LaNa_3BINOL_3$	$PrNa_3BINOL_3$	$NdNa_3BINOL_3$	$EuNa_3BINOL_3$	$YbNa_3BINOL_3$
$Ln-O_1$	2.423 Å	2.365 Å	2.338 Å	2.286 Å	2.203 Å
$Ln-O_2$	2.425 Å	2.386 Å	2.363 Å	2.312 Å	2.230 Å

Table 5.1: Ln-O distances in different $LnNa_3BINOL_3$ crystal complexes.

The possibility and the extent of axial accessibility to the metal is a matter of central importance in the description of the catalytic activity of these systems along th $4f$ period.

5.3   NMR data collection on $YbNa_3BINOL_3$

Peak assignments and relaxation data

The ^{1}H-NMR spectrum of the Yb^{3+} complex in d_8-THF solution is shown in figure 5.4. Six resonances can clearly be detected outside the usual *chemical shift* region (0-10 ppm). The number and the moderate linewidth of the signals allow us to assess that a single species is present in solution, stable and endowed with a D_3 symmetry on the NMR timescale. The assignment of the resonances has been performed through simple homonuclear decoupling experiments, since the line broadening is less than the coupling constant between protons.

The ^{13}C-NMR spectrum has been assigned, as far as the tertiary carbon resonances are concerned, on the basis of heteronuclear correlation experiments (HETCOR).

For each of the signals of the spectra, the values of the longitudinal and transverse relaxation times (T_1 and T_2) have been calculated; the results are displayed in table 5.2.

In the hypotesis of negligible contact contribution, the actual values of the pseudocontact shifts and dipolar relaxation enhancements employed in the structural optimization are calculated by subtracting shifts and relaxation rates of the corresponding corresponding Lu^{3+} complex. The ^{1}H-NMR spectrum of $LuNa_3(BINOL)_3$ is diplayed in figure 5.5; it has been assigned, together with the corresponding ^{13}C-NMR, with the same techniques discussed in the case of Yb^{3+}.

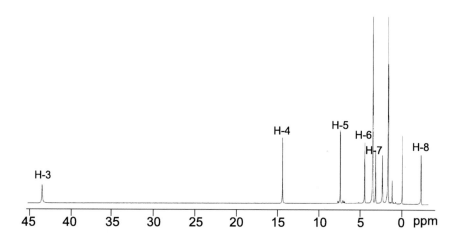

Figure 5.4: ^{1}H-NMR spectrum of YbNa$_3$BINOL$_3$ in d$_8$-THF.

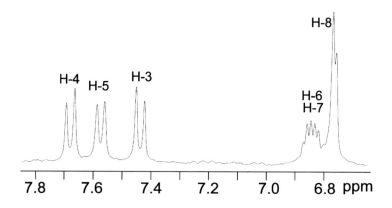

Figure 5.5: ^{1}H-NMR spectrum of LuNa$_3$BINOL$_3$ in d$_8$-THF.

Variable temperature spectra

The presence of a C$_3$ symmetric species in the crystals raises the question whether in solution an actual D$_3$ complex exhists or rather the higher symmetry in solution comes from an averaging process among two C$_3$ conformations (figure 5.6).

Variable temperature ^{1}H-NMR spectrum of YbNa$_3$BINOL$_3$ in d$_8$-THF were also recorded in the range $40°/-80°$C in order to monitor decoalescence of some signals following a possible symmetry reduction of the system.

Nevertheless, in the range examined, the only effect were the shift variation according to the predicted temperature dependence of the paramagnetic effect and a progressive broadening of the signals due to an increased Curie relaxation

pos.	^{1}H				^{13}C			
	δ Lu^{3+} (ppm)	δ Yb^{3+} (ppm)	δ^{pc} (ppm)	T$_1$ Yb^{3+} (ms)	δ Lu^{3+} (ppm)	δ Yb^{3+} (ppm)	δ^{pc} (ppm)	T$_1$ Yb^{3+} (ms)
2					164.1			
3	7.4	43.7	36.2	15.8	127.1	149.4	22.3	
4	7.6	14.5	6.9	137	128.3	132.7	4.4	234
5	7.5	7.4	-0.1	386	128.3	125.1	-3.2	335
6	6.8	4.5	-2.3	620	120.2	118.2	-2.0	181
7	6.8	3.2	-3.6	480	124.7	119.0	-5.7	286
8	6.7	-2.3	-9.0	133	125.9	117.2	-8.7	308

Table 5.2: NMR parameters of the YbNa$_3$BINOL$_3$ and LuNa$_3$BINOL$_3$ in d$_8$-THF at 25°.

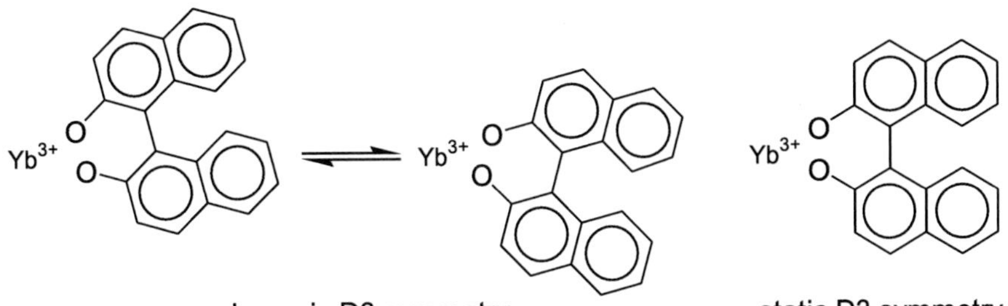

dynamic D3 symmetry   static D3 symmetry

Figure 5.6: Schematic representation of dynamic (left) and static (right) D$_3$ symmetry for YbNa$_3$(BINOL)$_3$. Only one of the three binaphthoate units is displayed.

contribution at lower temperatures.

5.3.1   Structural optimization

The data obtained from NMR were used as restraints in the structure determination of the paramagnetic complex in solution. Both in the case of C$_3$ and D$_3$ symmetries, for a given geometry surrounding the ion, the pseudocontact shifts depend only on one value $\mathcal{D}$ of the magnetic susceptibility (see eq. **??**). Furthermore, the NMR relaxation data allowed us to use in the fitting procedure additional constraints on the relative position of Yb^{3+} and the naphthyl moiety. The isolated evaluation of the Curie term according to eq. 2.58 of pag. 40, which relies on the difference between the reciprocal of the transverse and longitudinal relaxation times could not be employed due to errors in the evaluation of the T$_2$'s. For this determination, indeed, we used linewidths measurements, which suffer from field disomogeneity and from interference phenomena (transverse cross-correlation).

On the contrary, the generalized term of eq. 2.59 was employed for constrain the nuclear distances with respect to Yb(III). The constant C, which contains contributions from both the dipolar and the Curie relaxation, was fitted as well during the calculation (see section A.6.1).

On the basis of the collected NMR data, two different optimizations were performed, in order to take into account the two possible kinds (dynamic and effective) of D_3 symmetries, as depicted in figure 5.6.

- in the first case, the crystal structure was chosen as the starting geometry for the interpretation of the experimental data; a situation of dynamic D_3 symmetry (figure 5.6, a) was symulated by averaging between pairs of homolog nuclei A and B in the two naphthoate subunits of BINOL so that for a spin i the shift, relaxation time and distance r from the ion are given by:

$$\delta_i = \frac{1}{2}(\delta_A + \delta_B)$$

$$\frac{1}{T_{1i}} = \frac{1}{2}\left(\frac{1}{T_{1A}} + \frac{1}{T_{1B}}\right)$$

$$\frac{1}{r_i^6} = \frac{1}{2}\left(\frac{1}{r_A^6} + \frac{1}{r_B^6}\right)$$

During the optimization, conducted according to equations 2.23 and 2.59, the structure was kept fixed and only the $\mathcal{D} = \chi_{zz} - \frac{\chi_{xx} - \chi_{yy}}{2}$ value and the generalized relaxation constant C were allowed to change;

- in the second case, the crystal geometry was disregarded and all the degrees of freedom of the system were taken into consideration. Such a freedom is anyway reduced by the D_3 symmetry of the system (figure 5.6, b), so only 5 parameters have to be fitted, namely:

  1. the single $\mathcal{D}$ value (eq. ??) of the axially symmetric $\tilde{\chi}$ tensor (directed along the C_3 axis);

  2. the generalized relaxation constant C of equation 2.59;

  3. the Yb–O distance, d;

  4. the dihedral angle ϕ of the binaphthoate (figure 5.7);

  5. the dihedral angle α between the C_3 axis and the binaphthoate C1–C1' bond axis (figure 5.7).

The results of the optimization procedure are reported in table 5.3 and the values obtained through the fitting procedure are reported in table 5.4. The results obtained show that:

- the two models describe the same conformation of the binaphtoate moiety;

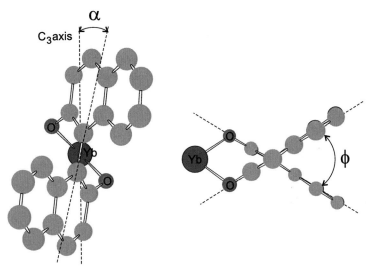

Figure 5.7: Dihedral angles ϕ and α in YbNa$_3$(BINOL)$_3$optimized in the simulation.

pos.	^{1}H					^{13}C				
	δ^{pc}_{exp} (ppm)	δ^{pc}_{dyn} (ppm)	δ^{pc}_{st} (ppm)	r_{dyn} (Å)	r_{st} (Å)	δ^{pc}_{exp} (ppm)	δ^{pc}_{dyn} (ppm)	δ^{pc}_{st} (ppm)	r_{dyn} (Å)	r_{st} (Å)
3	36.3	34.5	33.2	4.07	4.04	22.3	18.0	24.4	4.04	3.95
4	6.9	6.9	7.2	5.87	5.84	4.4	5.4	7.5	5.18	5.12
5	-0.1	-0.1	1.5	7.53	7.53	-3.2	-1.7	0.2	7.00	6.88
6	-2.3	-1.7	-0.9	8.59	8.59	-2.0	-2.8	-1.4	7.70	7.60
7	-3.6	-3.7	-3.4	7.95	7.95	-5.7	-5.3	-3.7	7.27	7.18
8	-9.0	-9.4	-9.6	5.98	6.03	-8.7	-8.9	-7.6	6.02	5.91

Table 5.3: Experimental and calculated shift values for the hydrogen and tertiary carbon nuclei after the optimization procedure, and distances from ytterbium. The values are reported both in the case of dynamical (*dyn* labelled) and static D_3 (*st* labelled) solution symmetry.

- only the relative position with respect to the ion is sligthly different in the two cases;

- the dynamic model derived from the crystal structure affords a better agreement factor in the fitting of both the experimental shifts and relaxation rates.

An analysis of the relative contribution of the S- and χ-mechanism to the nuclear relaxation (see section 2.2.3) can also be derived from the fitted constant C. The rotational correlation time τ_r can be estimated 130 ps from the ^{13}C-T_1

		DYNAMIC D₃-SYMMETRY	STATIC D₃-SYMMETRY
Yb–O distance	d_{Yb-O}	2.203 Å 2.230 Å	2.3 ± 0.2 Å
BINOL dihedral	ϕ	63°	62 ± 3 °
dihedral C$_3$ axis – C1/C1' bond	α	1.8°	5 ± 3 °
magnetic susceptibility ($ppm\cdot$ Å^3)	$\mathcal{D}$	1950 ± 120	1950 ± 120
relaxation constant ($\times 10^5$s Å^{-6})	C	2.83 ± 0.02	2.70 ± 0.02
agreement factor	R'	4.6	8.6
mean residual δ (ppm)	$\langle \delta^{pc}_{calc} - \delta^{pc}_{exp} \rangle$	1.4	2.0
mean residual T_1 (Å^{-6})	$\langle \frac{1}{r^6} - \frac{C}{T_1} \rangle$	0.28	0.42

Table 5.4: Results of the oprtimization procedure. Some structural parameters of the two geometric models employed are reported (see figure 5.7): in the DYNAMIC SYMMETRY column, the values are those of the crystal structure, while in the STATIC SYMMETRY column the numbers come from the optimization. The constants (magnetic susceptibility $\mathcal{D}$ and relaxation constants C) fitted, the agreement factor R' (as defined in section A.6.1) and the mean residuals are also displayed.

data of the diamagnetic Lu^{3+} complex according to the relation [11]:

$$\tau_r = \frac{T_1^{-1} \langle r_{HC} \rangle^6}{\left(\frac{\mu_0}{4\pi}\right)^2 \gamma_H^2 \gamma_C^2 \hbar^2} \tag{5.1}$$

Therefore, by the use of equation 2.60 of page 42, a value of about $1.3 \cdot 10^{-13}$ s for the τ_s of Yb^{3+} can be calculated, which is in total agreement with that found for the corresponding aquo complex and for other chelates [12, 13]. The relative contributions ρ of the Curie and dipolar relaxation terms in this systems can be estimated, according to equation 2.57, to be $\rho(\frac{Curie}{S}) \simeq 0.7$.

5.4   UV absorption and CD spectra

UV spectroscopy can provide valuable information on the structure of this kind of compounds. Of such an analysis, conducted elsewhere [14], the results that are relevant for the following will be here briefly summarized. Figure 5.8 (a) shows the UV absorbance spectrum of YbNa$_3$BINOL$_3$, compared with that of sodium 2-nafthoate. The agreement of positions and relative intensities of the bands allows one to approximate the electronic structure of the naphthalenic chromophores in YbNa$_3$BINOL$_3$ with that of 2-naphthoate; the apparent broadening of the most intense UV band in the spectrum of YbNa$_3$BINOL$_3$, associated to the $\pi \rightarrow \pi^*$ ^1B_b, is due to the splitting, excitonic in origin, of the corresponding band of the "monomer".

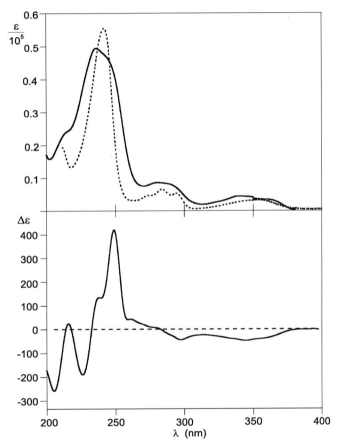

Figure 5.8: UV absorbance (a) and CD (b) of YbNa$_3$BINOL$_3$ in THF (c=0.5 mM, l=0.01 cm). In (a), the absorbance of 2-naphthol (1.07 mM in NaOH 0.1 M) is also reported (dashed line).

The corresponding CD spectrum (figure 5.8 (b)) features a *couplet* with negative component at 226 nm and positive at 260 nm, due to the coupling of the ^1B_b transitions of the naphthoate moieties. It must be noted that the apparence of both spectra, with narrow and well-resolved transitions, is likely to witness the presence of a single species in solution.

The magnitude of the apparent splitting between the two bands of the *couplet* is related to the dihedral angles between the transition dipole moments of the chromophore units [15].
Assuming that:

- the geometry of the crystal structure is conserved in solution in agreement with what discussed above;

- the auxochrome group $-O^-$ induces a deviation of about $15°$ of the ^1B_b transition moments from the direction of the aromatic long axis (as suggested by CNDO-CI calculations),

the DeVoe method [16, 17] can be adopted for the simulation of the CD pattern. Taking into account both inter- and inter-binaphthylic couplings, a $\Delta\lambda_{max}$ of 27 nm is predicted, in great agreement with the experimental datum.
The spectra of the analogous lutetium derivative are identical, proving the isostructurality of the two species, which had been assumed using $LuNa_3BINOL_3$ as a diamagnetic reference in the NMR analysis.

5.4.1   Effect of the alkaline ion: data on YbK_3BINOL_3

In order to get insight about the role of the second cation component of the complex, the analogous potassium derivative YbK_3BINOL_3 was synthetised.
The UV aborbance and CD spectra of YbK_3BINOL_3 are perfectly superimposable to those of $YbNa_3BINOL_3$ [14], pointing that no big changes (if any) occur both in the binaphthoate angles and in the relative disposition of the three BINOL units. The ^{1}H-NMR spectrum of YbK_3BINOL_3 (figure 5.9), though, is deeply varied upon the replacement of Na with K: the whole spectral pattern spans now a reduced window, the most shifted signal resonating now at about 20 ppm.
After a closer inspection of the shifts, however, we can notice that the relative position of the signals are the same in the two spectra. After substraction of the diamagnetic shifts of $LuNa_3BINOL_3$, the two sets of shifts are exactly proportional. Such a result is analogous to that found in section 3.4.2 (page 71) for the spectra of YbDOPEA and YbDONEA (Yb-(R)-**3** and Yb-(S)-**4**) in different solvents.
The different shifts foreseen by eq. **??** are therefore not caused by different geometrical factors (i.e.: different positions of the nuclei with respect to Yb^{3+}), but are due to variations in the $\mathcal{D}$ parameter of the ion (from about 2000 in the Na derivative to about 1300 in the K complex).
The picture is in agreement with the UV-CD data and indicates that no major differences concern the geometry of the molecule. The origin of the difference in the $\mathcal{D}$ parameter must be looked for in the different crystal field experienced by the lanthanide ion. According to Bleaney's theory (see section 2.1.3), the $\mathcal{D}$ factor is proportional to the crystal field coefficient $B_0^2 = \langle r^2 \rangle A_0^2$ (see pages

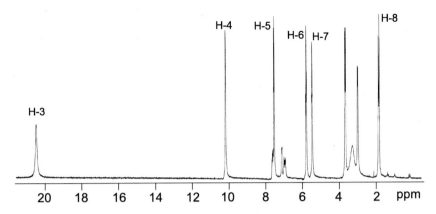

Figure 5.9: ^{1}H-NMR spectrum of YbK$_3$BINOL$_3$ in d$_8$-THF.

?? and 7), which in turn is dependent – to the first order – on the charge geometry surrounding the ion (see section 1.1.3):

$$\mathcal{D} \propto -B_0^2 \propto -\sum_L q_L \frac{3\cos^2\theta_L - 1}{r_L^3} \tag{5.2}$$

where the sum is extended to all the ligands L, whose positions with respect to the ion principal axes system are defined by θ_L and r_L.
Two main factors can result in a decrease of the $\mathcal{D}$ factor and can be both operative in the present case:

- a reduction of the negative charge on the naphtholate oxygens (for which $57° < \theta < 90°$ and $3\cos^2\theta - 1 < 0$) in the K derivative respect to the Na molecule;

- an increased H$_2$O axial coordination ($\theta = 0$, $3\cos^2\theta - 1 > 0$), which can also be a result of the effect described in the first point.

An estimate of the degree of axial coordination can indeed be obtained by the water signal, which can shifted up to 5 ppm in the spectrum of YbK$_3$BINOL$_3$, depending on the water content of the sample. Assuming a bond distance r of 2.6 Å, as for example in the crystal structure of Yb-**3**, axial position of the bond ($\theta = 0$) and a magnetic susceptibility $\mathcal{D}$ of 1300 ppm/Å^3, the chemical shift of the bound water could be calculated to be:

$$\delta_w^{bound} = \mathcal{D}\frac{3\cos^2\theta - 1}{r^3} \simeq 150ppm$$

so that the observed shift should withness of a 2% bound water.
Further evidence of the variation of the crystal field experienced by the Yb^{3+} in the two different complexes will be found in the picture of the electronic structure of the ion provided by NIR spectroscopy.

5.5   NIR absorption and CD spectra

The near infrared absorption spectra of $YbNa_3(BINOL)_3$ and YbK_3BINOL_3 in THF solutions are shown in figure 5.10.

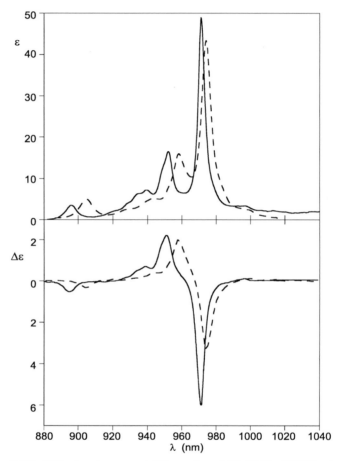

Figure 5.10: NIR absorbance and CD spectra of (S)-$YbNa_3BINOL_3$ (continuous line) and (S)-YbK_3BINOL_3 (dashed line) in d_8-THF solution (c=5mM and 19mM, respectively; l=1cm).

As discussed in section 1.4.3, ytterbium $^2F_{7/2} \rightarrow {}^2F_{5/2}$ transition is split by a C_3 symmetric crystal field in 12 components around 980 nm. In the present case, the separation of the levels is supposed to be particularly large, since a band can be found even below 900 nm, almost 100 nm less than the fundamental of the free Yb^{3+}. As the energy spacing between the crystal field sublevels gets larger, the Boltzmann population of the higher levels gets smaller and not all the 12 transitions will probably be detectable.

Only three main transitions can be easily revealed indeed and they are all quite narrow (5 nm half-height linewidth): a very intense signal at 974 nm (in both species, a weaker one at 955 nm (958 nm for YbK_3BINOL_3) with two shoulders in the 935-945 region and a smaller signal at 895 nm (905 nm for YbK_3BINOL_3) . A very pale and broad transition is finally detectable at about 1000 nm.

Correspondingly, the CD spectrum in the same region (see figure 5.10, bottom) show a sequence of alternating bands (with some shoulders) which, within the experimental error, arise at the same frequencies of the main absorption bands. The intensity of the dichroic signals is surprisingly high and, for the transition at 974 nm in $YbNa_3(BINOL)_3$, we can calculate a dissymmetry factor $g = \Delta\epsilon/\epsilon$ of about 0.12, which witnesses a very efficient occurrence of asymmetric induction from the binaphthoate ligands to the electronic levels of the lanthanide.

It is noteworthy that the shape of the spectra of the two complexes examined is the same, the only difference being the range of wavelengths in which the transitions are spread. This conclusion matches perfectly the NMR counterpart, once again showing larger spectral widths (as a consequence of larger crystal field parameters) for ions in complexes with larger magnetic anisotropies.

Even if a number of transitions are missing, thus preventing a thorough quantitative analysis of the energy level splitting and the evaluation of the crystal field parameters, the present is one of the few cases of a NIR-CD spectrum of lanthanide compounds whose solution geometry is known and relatively simple and can represent an interesting reference point for the developement of empirical correlations involving a wider set of molecules.

5.6   Conclusions

In this chapter two ytterbium-based catalysts were synthetised and their structure was cross-investigated both in the solid state and in solution. The X-ray analysis of $YbNa_3BINOL_3$ showed that this molecule possesses a structure very similar to the other lanthanide derivatives previously characterized, the only differences coming from the bond distances, shorter in the present complex. The shrinkage of the coordination polyhedron causes as a result a reduction of the coordination number, since the water molecules which axially approaches the central cation is absent in the ytterbium complex. Analysis of the paramagnetic shifts and relaxation enhancements showed how the C_3 symmetry of the crystal state averages to a D_3 symmetry in solution thank to a rapid interchange of two different C_3 conicities, whose structure is compatible with that determined in the solid state. No evidence of axial water coordination was collected in the soluiton study. YbK_3BINOL_3 is fully isostructural to the Na analogue as far as the ligand positions and conformations are concerned. The most electronegative co-cation, anyway, causes a variation in the naphtholate oxygen charge. As a consequence, the crystal field experienced by the lanthanide is different and also its axial accessibility is increased. It is interesting

to note that CD spectroscopy proved a particularly versatile technique, able
to consistently describe both steps of this phenomenon: the spectra of the two
molecules are identical in the UV region, but the differences concerning the
first coordination sphere show up just switching to upper wavelengths.

The results of the present chapter, beside being of relevance for the understanding of the catalytic mechanism of this class of molecules, are a wonderful
example of how successful the combination of different techniques can be in
the investigation of solution structural features.

Bibliography

[1] KOBAYASHI, S., *Lanthanides: Chemistry and Use in Organic Synthesis;*
    SPRINGER-VERLAG: BERLIN, 1999.

[2] ANWANDER, R., *Top. Curr. Chem.* **1996,** *179,* 33–112.

[3] ANWANDER, R., *Top. Curr. Chem.* **1996,** *179,* 149.

[4] SHIBASAKI, M.; SASAI, H.; ARAI, T., *Angew. Chem. Int. Ed. Engl.*
    1997, *36,* 1236–1256.

[5] SASAI, H.; ARAI, T.; SATOW, Y.; HOUK, N. K.; SHIBASAKI, M.,
    J. Am. Chem. Soc. **1995,** *117,* 6194–6198.

[6] SASAI, H.; SUZUKI, T.; ITOH, N.; TANAKA, K.; DATE, T.;
    OHAMURA, K.; SHIBASAKI, M., *J. Am. Chem. Soc.* **1993,** *115,* 10372–
    10373.

[7] SASAI, H.; WATANABE, S.; SHIBASAKI, M., *Enantiomer* **1997,** *2,*
    267–271.

[8] GRÖGER, H.; SAIDA, Y.; SASAI, H.; YAMAGUCHI, K.; MARTENS, J.;
    SHIBASAKI, M., *J. Am. Chem. Soc.* **1998,** *120,* 3089–3103.

[9] ABRAM, U.; DELL'AMICO, D. B.; CALDERAZZO, F.; PORTA, C. D.;
    ENGLERT, U.; MARCHETTI, F.; MERIGO, A., *Chem. Commun.* **1999,**
    2053–2054.

[10] DICKINS, R. S.; HOWARD, J. A. K.; MAUPIN, C. L.;
     MOLONEY, J. M.; PARKER, D.; RIEHL, J. P.; SILIGARDI, G.;
     WILLIAMS, J. A. G., *Chem. Eur. J.* **1999,** *5,* 1095–1105.

[11] ABRAGAM, A., *Principles of nuclear magnetism;* OXFORD UNIVERSITY
     PRESS: LONDON, 1961.

[12] ALSAADI, B. M.; ROSSOTTI, F. J. C.; WILLIAMS, R. J. P., *J. Chem.*
     Soc., Dalton Trans. **1980,** 2147.

[13] PETERS, J. A.; HUSKENS, J.; RABER, D. J., *Progr. NMR Spectrosc.*
     1996, *28,* 283–350.

[14] PESCITELLI, G., PH.D. THESIS, UNIVERSITY OF PISA, 2001.

[15] DI BARI, L.; PESCITELLI, G.; SALVADORI, P., *J. Am. Chem. Soc.*
     1999, *121,* 7998–8004.

[16] DeVoe, H., *J. Chem. Phys.* **1964,** *41,* 393.
[17] DeVoe, H., *J. Chem. Phys.* **1965,** *43,* 3199.

Part III

Labile lanthanide adducts

Chapter 6

Spectroscopic determination of the configuration of chiral diols

Chiral diols are ubiquitous in natural compounds and are largely used as chiral building blocks in organic chemistry. Facile methods of enantioselective dihydroxilation of olefins make them readily accessible at high enantiomeric purities also through synthesis.

Unfortunately, predicting their absolute configuration may not be trivial. To this end, CD spectroscopy is one of the most powerful techniques: in the presence of cyclic structures, non-empirical exciton-coupling (ECCD) of dibenzoates or analogous derivatives can be used.

Furthermore, diaryl ethandiols can be made rigid through formation of a ketal, thus opening again the way to ECCD in substrates endowed with suitable chromophores. In general, however, the use of exciton coupling is suitable for molecules with restrained conformational freedom and with strong electronic absorptions. Since the glycol moiety is practically transparent down to 190 nm, observation of a CD spectrum relies on the formation of cottonogenic derivatives, *e.g.* through chelates of various transition metal systems. For example, Snatzke demonstrated that following interaction of a chiral diol with dimolybdenum tetraacetate, circular dichroism in the electronic transitions of the inorganic complex is generated, which can be used for probing the configuration of the diol. This method is very simple and rather versatile. The main drawbacks can be ascribed to the time-evolution of the signals, not yet clarified, and to possible interferences due to chromophores absorbing above 280 nm possibly attached to the diol.

Lanthanide ions lend themselves as promising alternatives, owing to the high affinity for oxygen ligands and to the possibility to expand their coordination number.

6.1   The lanthanide diketonate method

6.1.1   Generalities

In 1971 Koji Nakanishi reported with James Dillon a new method for assigning the absolute configuration of chiral diols by the observation of induced CD on lanthanide diketonates [1, 2]. Those days several metal complexes had already been tried (and many more had to come, yet), exploiting the chelating ability of the glycols; two opposite attitudes followed: observing dichroism on the metal-centred transitions like in the methods of cuprammonium [3–5], osmium [6] and molybdenum [7, 8], just to quote a few examples; or looking at the CD of the ligand absorptions, which implies the use of chromophoric ligands, such as among the others, Ni(acac)$_2$ and Pr(dpm)$_3$[1] [9].

In a very comprehensive study published in 1975 [10, 11], application of the latter two systems is critically investigated. In principle the method is very simple, consisting in the mixing of diol and metal complex and observing a neat CD couplet at about 300 nm. In spite of this attractive feature, unfortunately, severe precautions must be taken, which limits the scope and often precludes practical applications. Indeed, reports of the use of this methods are scarce.

That the matter is still of great scientific interest is demonstrated by the progress in the preparation of 1,2-diols by enantioselective olefin dihydroxilation [12], which opened the way to a larger use of these molecules as chiral building blocks in more complicated architectures. Sharpless reaction is associated with a mnemonic rule predicting its stereochemical outcome, but this lies on an empirical ground and some exceptions have been reported [13–22]. Notably, when applied to chiral substrates, double asymmetric induction [23] (from the substrate and from the auxiliary) may lead to products of unpredictable configuration [24, 25] and a direct confirmation of the stereochemstry is required.

A few cases were examined by means of a variation of Dillon and Nakanishis method, using Eu(fod)$_3$ - a fluorinated relative to Pr(dpm)$_3$, and a popular NMR shift reagent. Partridge, observed that Eu(fod)$_3$ gives rise to stronger and stable (see below) CD signals when interacting with a variety of cholestan pentols [26–30], and confirmed the results by diffractometric analysis. Surprisingly, to our knowledge, no comprehensive and systematic study followed this first report.

The same reagent, together with other lanthanide analogues was very recently used by Tsukube *and coll.* [31–33] for the analysis of aminoalcohols[2].

Yb(fod)$_3$ (figure 6.1) was mixed in CCl$_4$ solutions with several different diols, studying the mixture by means of CD in the UV and in the near infrared (NIR, in the range 900-1000 nm, where Yb^{3+} has its ff transition) and by NMR spectroscopy.

---

[1]For all the avvreviations, see Appendix C.

[2]Unlike what demonstrated in the present chapter, these researchers were not able to detect induced CD with diols.

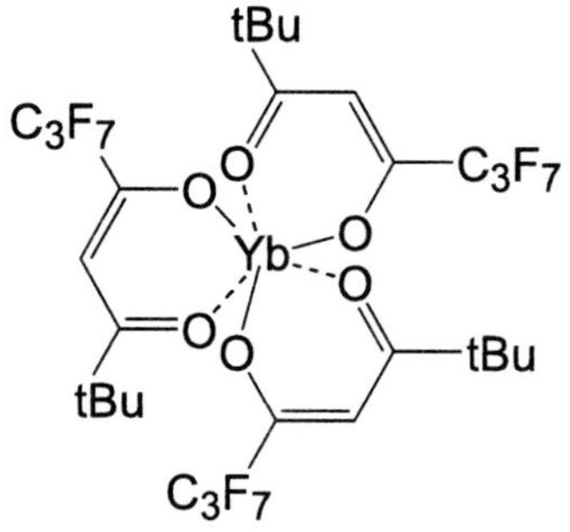

Yb(fod)₃ **12**

Figure 6.1: Yb(fod)₃.

The resulting picture is fairly complex, but most drawbacks of Pr(dpm)₃ are overcome by Yb(fod)₃, which can thus be regarded as a useful tool for assigning the absolute configuration of diols.

One particular merit of this reagent consists in the possibility of using **two spectral windows** for chiroptical measurements: in the UV and in the NIR: in the UV there is a great sensitivity, which allows one to use minute amounts of analyte; on the other hand, for strongly UV-absorbing diols interference effects may prevent a correct analysis in that region; NIR observation overcomes the problem, switching to wavelengths where virtually no organic substituents can contribute and where Yb^{3+} has a CD-sensitive transition charaterized by a very favourable g-factor (see section 1.2.4). Moreover, as it will be demonstrated in the following, it yield unique information on the nature of the chiral species.

The structure and the dynamics of the optically active species can be thoroughly investigated by means of the concerted use of ¹H-NMR of the paramagnetic complex and of chiroptical measurements. Although the process of chiral induction of the diol onto the Yb^{3+} complex cannot be rationalized in terms of simple sterical arguments, the spectroscopic data point toward a clear structural picture.

The configurational inference obtained by this method comes up to other similar ones, most notably that based on molybdenum tetracetate [7, 8, 34].

6.1.2   A survey of the diketonates method and its extention to the NIR

The method set up by Dillon and Nakanishi consists in mixing a slight excess (1 to 4 equivalents) of a chiral diol with Pr(dpm)₃ in an apolar solvent, as CCl₄, chloroform or hexane.

The authors observe a strong competition with ethanol (often contained as a stabilizer) and water, which makes mandatory the use of carefully purified and dehydrated solvent, reagent and analyte. Soon after the mixing, a CD spectrum in the region 250-350 nm is observed, in the form of a more or less regular couplet. The amplitude of the signals is reported strongly time-dependent. Upon changing the molar ratio between analyte and auxiliary, to a first increase of intensity, follows a steady reduction.

All these points indicate complex solution equilibria, where the lanthanide is likely to coordinate eventually more than one diol molecule, residual water, as well as to possibly loose a diketonate ligand.

In order to get a feeling about the use of a highly symmetric Yb diketonate, $Yb(dpm)_3$, ICD spectra were recorded in the NIR after addition of chiral diols. Indeed, soon after the mixing a strong Cotton effect centred at about 976 nm can be observed, dominating over much smaller features. This dichroic band is strongly time dependent in a rather unpredictable fashion: it is apparently influenced by the exact composition of the solution as well as by temperature [3]. Clearly, there are different species in solution which interconvert on the minute timescale. Moreover, magnitude and sign of the NIR-CD can be affected by the amount of residual water. Thus, the results could hardly be rationalized. This is more or less in-keeping with what observed for the UV-CD.

The origins of the ICD spectra in the two wavelength regions, reside in very different processes. At higher energy, in the UV, it is associated to the electronic transitions localized on the diketonates.

Circular dichroism can arise from the exciton coupling between the electric dipole allowed transitions located on the diketonates. Indeed if for d-metals acetyacetonates, like Ni and $Co(acac)_3$, a charge transfer band is also visible, in the present case this is unlikely, owing to the larger energy splitting between ligand and metal transitions.

The analysis of the shape of the UV-CD spectrum should in principle provide direct structural information on the chiral species which is formed upon incorporation of the diol. However, the resulting complex, containing three diketonates and one (possibly C_2 symmetric) diol, must have a low symmetry (this point will more deeply discussed in the following). Thus all possible pairwise interactions between the electronic transitions on the diketonates should be taken into account, which hampers a correct non-empirical analysis by means of simple structural arguments.

In the NIR, instead, it is the chiral environment around the metal ion which causes the appearence of a non-vanishing rotational strength in the magnetic-dipole ff transition multiplet of ytterbium. To-date extension to ff transitions, especially in the NIR of the configurational/spectroscopic correlation using diketonates seems to be lacking.

---

[3]This is in spite of using carefully purified and dehydrated solvent and reagents.

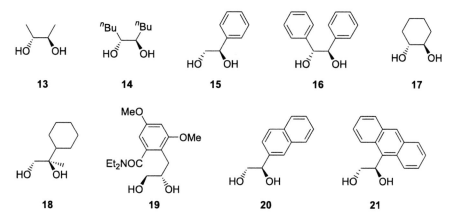

Figure 6.2: Chiral 1,2-diols employed for the analysis.

6.1.3 UV-CD spectra of mixtures of chiral diols and Yb(fod)$_3$

Mixtures of Yb(fod)$_3$ with chiral diols **13–21** (see figure 6.2) in about 1:1 ratio in CCl$_4$ or chloroform typically yield the UV-ICD spectra shown in figure 6.3. They are undefinitely stable (over several days), as already observed by Partridge [26, 27]. Moreover, upon addition of water, there is an intensity decrease but the shape and the sign of the the spectrum remains unaltered.

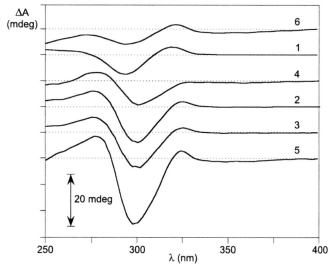

Figure 6.3: UV-CD spectra of diols **13–18**, with Yb(fod)$_3$, molar ratio 1:1, in anhydrous CCl$_4$ 0.35 mM. Cell lenght 0.1 cm, room temperature.

Unlike what reported for mixtures of Yb(fod)$_3$ with aminoalcohols [31–33] and for the ICD of Ln(dpm)$_3$ [1, 2, 11], the spectra of figure 6.3 are strongly asymmetric and several components are clearly detectable instead of two bands of opposite rotational strengths. Owing to the low symmetry, in the chiral species the three diketonates are not equivalent and must give rise to different pairwise interactions that can in principle account for the asymmetric spectrum.

The overall appearence of the spectra is slightly substrate-dependent, with more or less pronounced smaller features, such as the two positive bands flanking the strongest negative Cotton effect at 300 nm. For **13**, there is a slight blue shift of all dichroic lines.

Nonetheless, in this case, as well, at least three Cotton effects of alternated sign can be detected. In spite of these minor variations, the sign of the band at 300 nm appears perfectly correlated with the absolute configuration of the diol, being negative for all (R) or (R, R) substrates. This is remarkable considering the structural differences between the molecules investigated. As seen in other similar cases, there is a variation in the intensity of the spectra, which may partly reflect minor differences in the stability constants of the chiral complexes, or a different ability in the chiral induction mechanism.

The role of the molar ratio between diol and Yb(fod)$_3$ was investigated for the case of the simplest term: (R,R)-2,3-butanediol **13**, as described below.

6.1.4   NIR-CD spectroscopy

The NIR-CD spectra shown in figure 6.4 reveal a good correlation between absolute configuration and the sign of a strong dichroic band at 976 nm. Thus (R) or (R,R)-diols feature a negative Cotton effect. There are, however, two notable exceptions, (R,R)-2,3-butandiol **13** and (R,R)-5,6-decandiol **14**, which give a positive, dissignated band at 978 nm.

Extension of the spectral observation to the NIR opens the way to the analysis of substrates containing strong UV-Vis chromophores, which introduce more or less strong distortions in the CD spectra obtained by other techniques. Thus, for example a neat NIR-CD spectrum is obtained for the mixture, (R)–1–(9–anthracenyl)–1,2–ethanediol **21**/Yb(fod)$_3$, whereupon the strong absorption bands of the condensed aromatic group messes up the UV-CD data.

It is noteworthy that this molecule cannot be treated with other very successful methods, like the one based on Mo$_2$(AcO)$_4$ [7, 8].

6.1.5   NMR of 1:1 mixtures of diols and Yb(fod)$_3$

The proton spectrum of Yb(fod)$_3$ in CDCl$_3$ reveals two lines (see Table 1) which can be assigned to the ^{t}Bu group and the CH, on the basis of the relative integrals. The existence of one signal for each proton type demonstrates an effective axial C$_3$ symmetry of the molecule [4], which may arise from dynamic

---

[4]The effective C$_3$ symmetry is further supported by the ^{19}F-NMR spectrum, where only three lines can be detected, in agreement with the perfluoropropyl chain.

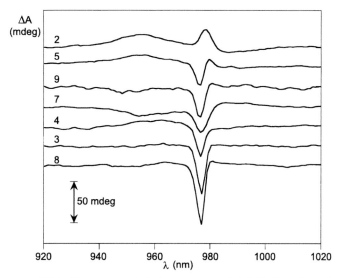

Figure 6.4: NIR-CD spectra of diols **13–21**, with Yb(fod)$_3$. Path length 10 cm, 5 mM. In order to favour the comparison, the spectrum of diol **19**, having opposite configuration, was inverted, while those of molecules **14**, **17** and **21** have been multiplied by a factor of 2.

exchange of the ligand among non-equivalent sites. For a sistem with this symmetry the pseudocontact interaction has the very simple expression of equation 3.2.

A further very broad peak is found at 203 ppm, assigned to a bound water molecule, demonstrating that in Yb(fod)$_3$ ytterbium is seven-coordinated in chloroform solution. The absence of the typical resonance at 1.65 ppm for water indicates that there is fast exchange between free and bound forms.

Through a titration of a solution of Yb(fod)$_3$ with water, the limiting paramagnetic shift can be estimated about 250 ppm and, in a mM-range solution, water is bound to over 8 %. Assuming that bound water molecule is along the principal axis of the magnetic susceptibility anisotropy tensor and that the distance Yb–O is 2.6 Å, one can obtain a rough estimate of the principal component of the tensor, $\mathcal{D} \simeq 3500$ ppm Å^3. After addition of one equimolar quantity of **13**, the resonances listed in Table 6.1 can be detected.

Each proton can be assigned on the basis of the integrals (for the CH$_3$) and of the relative linewidth: because the OH is nearer to the paramagnetic centre, it is expected to be more effectively relaxed and it is furthermore involved into exchange processes; both mechanisms concur in broadening its resonance with respect to the methyne proton.

The position and the number of the diol signals indicate that:

- in the presence of the chiral diol, a new paramagnetic species is created;

- free and bound diol are in fast exchange;

		diol			Yb(fod)$_3$	
(R, R)–2,3–butandiol	H$_2$O	OH	CH	CH$_3$	CH	^{t}Bu
free	-	3.9	3.5	1.15	-	-
Yb(fod)$_3$ 1:1	25.0	68.5	56.4	23.3	1.8	-15.4

Table 6.1: ^{1}H–NMR shifts (ppm) of a solution of free (R,R)–2,3–butandiol **13** and of a 1:1 mixture with Yb(fod)$_3$ (anhydrous CDCl$_3$, [**13**]= 0.10 M).

- there is a dynamic symmetrization of the adduct.

6.1.6 Variable molar ratio data

A full titration with **13** was performed, followed by means of all the spectro-scopic techniques mentioned above. There is a variation of the appearence of the CD spectra as depicted in figure 6.5, which demonstrates the existence of several equilibria between different species in solution.

The CD in the NIR clearly demonstrates the existence of two chiral species, withnessed *inter alia* by a progressive appearence of a negative Cotton effect at 975 nm on increasing the diol-to-metal ratio. The CD in the NIR clearly demonstrates the existence of two chiral species, witnessed inter alia by the progressive appearence of a negative Cotton effect at 975 nm on increasing the diol-to-metal ratio. This phenomenon is easily overlooked by UV-CD investi-gation alone, where the spectral changes are less striking. Exactly the same kind of evolution of the NIR CD is found also on addition of a monodentate alcohol (EtOH or BuOH) to a 1:1 mixture of diol **13** and Yb(fod)$_3$. This indi-cates that a diol may enter with two molecules in the coordination sphere of the lanthanide, once as a chelating agent, once as a monodentate ligand. The nature of the chiral species is best investigated by the quantitative analysis of the ^{1}H-NMR titration data shown in figure 6.6.

A small amount of residual water (essentially due to the hygroscopic lanthanide complex) is initially bound to Yb(fod)$_3$. From the sign of the lanthanide-induced shifts (see also the discussion in the next section), one detects that it must occupy an axial position, yielding a 7 or maybe 8-coordinated complex (mono- or bi-capped trigonal antiprismatic). On adding **13**, water is eventually displaced by the diol. The fod resonances evolute reaching a plateau about at 2:1 mixture; different processes may account for this behaviour:

- a structural rearrangement, taking the diketonates to a different position with respect to the principal axis of the magnetic anisotropy;

- a variation of the anisotropy parameter $\mathcal{D}$, following complexation with the diol;

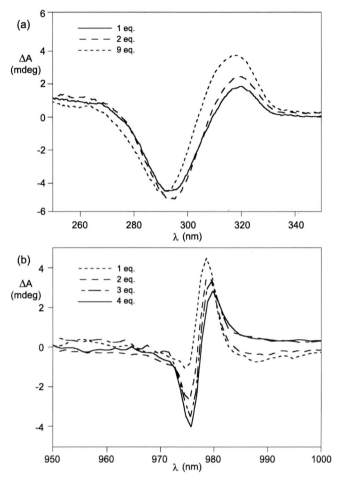

Figure 6.5: a) UV-CD spectra at different molar ratio of (R, R)–2,3–butanediol **13**. Solvent: anhydrous CCl_4, path lenght 0.1 cm, 0.18 mM. b) NIR-CD spectra of $Yb(fod)_3$ at different molar ratio of (R, R)–2,3–butanediol **13**. Solvent: anhydrous CCl_4, path lenght 10 cm, 4.8 mM.

- the release of a molecule of diketonate, in such a way that free and bound ligands are in fast exchange and the observed shift is a weighted average of the two limiting values.

The first two processes must both occur and contribute to some extent, because the chiral complex is globally different from free $Yb(fod)_3$ (more or less bound to water): indeed, the role of the axial substituent to the observed magnetic anisotropy has been demonstrated (see section 3.6).

The third hypothesis can be ruled out, on the basis of the the asymptotic paramagnetic shifts of fod: if one diketonate did leave the paramagnetic complex,

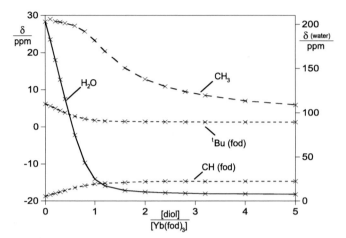

Figure 6.6: ^{1}H–NMR shifts of selected protons as a function of diol-to-Yb(fod)$_3$ molar ratio. The relative error on the shifts is essentially due to the linewidth and it is always below 2 %.

the observed shift would be the average:

$$\delta_{obs} = 2 \cdot \delta_{bound} + \delta_{free}$$

which leads to unacceptable estimates for δ_{bound} of both methyne and ^{t}Bu protons.

In a 1:1 diol-to-metal mixture, taking into account the presence of residual water at least three equilibria may be considered:

$$Yb(fod)_3 + H_2O \rightleftharpoons Yb(fod)_3 \cdot H_2O$$
$$Yb(fod)_3 \cdot H_2O + \mathbf{13} \rightleftharpoons Yb(fod)_3 \cdot \mathbf{13} \cdot H_2O$$
$$Yb(fod)_3 \cdot H_2O + \mathbf{13} \rightleftharpoons Yb(fod)_3 \cdot \mathbf{13} + H_2O$$

The former describes dissociation of water from Yb(fod)$_3$; it does not explicitly involve the diol and has been already discussed previously. The latter two lead to the formation of chiral Yb species: one is an associative process whereupon Yb becomes 9-coordinated, the other is a ligand exchange between water and diol. Naturally this is only one of the possible representations of the processes. An apparent affinity constant between **13** and Yb(fod)$_3$ can be calculated by a quantitative analysis of the NMR data and is $7(\pm 1)$ 10^2 M^{-1}. A detailed discussion of the structure of the chiral adduct(s) wil be presented in the following sections. A further diol molecule or a monodentate alcohol may play the role of water in the associative process, leading to a different chiral complex, as represented by the equation

$$Yb(fod)_3 \cdot \mathbf{13} \rightleftharpoons Yb(fod)_3 \cdot \mathbf{13} \cdot ROH$$

(where ROH represents an alcohol molecule or the diol acting as a monodentate ligand). It is noteworthy that the formation constants for the other diols is of the same order of magnitude as demonstrated by the similarity of the observed shifts for all the substrate protons [35] (see table 6.2).

	H_a, H_b	H_c
(R,R)-2,3-butandiol (**13**)	56.4	68.5
(R,R)-5,6-decandiol (**14**)	67.7	81.5
(R)-1-phenyl-1,2-ethandiol (**15**)	73.9, 75.6	93.7
(R)-1-(2-naphthalenyl)-1,2-ethanediol (**20**)	53.2, 58.9	73.2

Table 6.2: ^{1}H–NMR shifts (ppm) for different diols with Yb(fod)$_3$, ratio 1:1, 0.1 M, anhydrous CDCl$_3$.

6.1.7   Analysis of the NMR pseudocontact shifts

Analysis of the ytterbium–induced shifts experienced by the diol can provide a picture of the complexation and some accounts on the interaction geometries between the shift reagent and the substrate [36].

The lanthanide–induced shifts for the three H atoms of the diol **13** in the 1:1 complex were extrapolated from the data of the incremental addition of diol to Yb(fod)$_3$: the experimental pseudocontact terms δ_{exp}^{pc} for the bound form were then corrected using the corresponding values of the free diol as diamagnetic references and are reported in Table 6.3.

		g^- conformation			g^+ conformation		
proton	δ_{exp}^{pc}	$\left\langle \dfrac{3cos^2\theta - 1}{r^3} \right\rangle$	δ_{calc}^{pc}	$\delta_{exp}^{pc} - \delta_{calc}^{pc}$	$\left\langle \dfrac{3cos^2\theta - 1}{r^3} \right\rangle$	δ_{calc}^{pc}	$\delta_{exp}^{pc} - \delta_{calc}^{pc}$
CH	68.4	0.032	69.5	-1.2	0.022	64.4	4.0
CH$_3$	27.9	0.012	25.1	2.8	0.013	36.0	-8.2
OH	83.7	–	–	–	–	–	–
$\mathcal{D}$ (ppm Å^3)		2170 ± 90			2870 ± 350		
R (%)		4.1			12.3		

Table 6.3: Experimental δ_{exp}^{pc} and calculated δ_{calc}^{pc} pseudocontact shifts for the protons of diol (**13**) complexed with Yb(fod)$_3$. Two possible g^- and g^+ conformation of the diol are considered. The fitting procedure has been performed on the CH and CH$_3$ protons only.

The chemical shifts of the diketonate units cannot be exploited in a quantitative fashion, since they are average values of dynamically symmetrized species. Anyway, we must observe that their value does not sensibly change and, what is more important, their sign is unaffected after the diol complexation and it is opposite to that of the diol protons.

Therefore, the diketonates are arranged equatorially around the metal ion, while the diol molecule binds along one of the two axial cones defined by eq. 3.2, that is along the effective C_3 axis. The picture that one can derive from these data is that the chiral species is highly fluxional but the average positions of diol and fod relative to Yb are well determined and can be depicted as in figure 6.7.

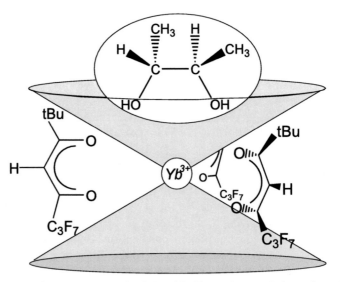

Figure 6.7: Schematic model of the (R, R)–2,3–butanediol **13** bound to a molecule of Yb(fod)$_3$. Proton inside the double cone are downfield shifted, outside these cones the paramagnetic shifts are upfield. This analysis does not allow to determine if the ^{t}Bu and the C$_3$F$_7$ groups are actually oriented as depicted.

The situation is analogue for all other diols, for which the shifts after addition of Yb(fod)$_3$ are invariably upfield (between 5 and 100 ppm). It is noteworthy that the observed signals for the hydroxyl and carbinol protons in all 1:1 mixtures resonate at very similar values.

The analysis can go further. Two distinct model structures for the chiral species can be built optimizing with MM2 calculations a g^+ and a g^- conformation of **13** (figure 6.8) and locating them symmetrically along the C_3 axis of Yb(fod)$_3$; bonding distances of 2.5 Å between the lanthanide and the oxygens were derived from data available on a large set of ytterbium compounds [37]. Averaged geometric parameters were calculated for the CH and methyl protons in the two reference models and are displayed in Table 6.3 as well.

The two pseudocontact chemical shifts were then fitted against the geometrical factors according to eq. 3.2. The results of such procedure are also reported

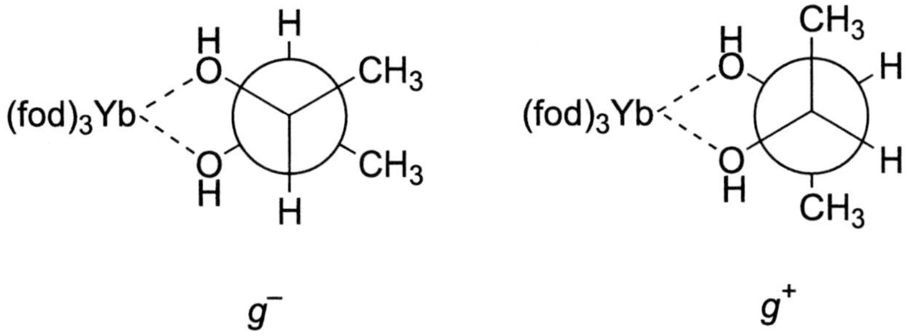

Figure 6.8: Different chelations following a g^+ or a g^- conformation of butanediol **13**.

in Table 6.3.

In the case of the g^- conformation, the agreement factor R between experimental and calculated values is definively better, showing that this is the species binding Yb in solution and confirming that the steric hindrance of the substituents in the diol determines the chirality of the chelation.

The value of the magnetic anisotropy parameter $\mathcal{D}$ is strongly reduced after the diol complexation. It must be observed that such a behaviour relates with the perturbation of the crystal field around the ion brought about by the different substituents.

Even if a structural rearrangement of the diketonate moieties cannot be excluded, the substitution of a water molecule with the bidentate diol in axial position, as well as the addition of the diol to the hydrated complex, can totally account for the observed phenomenon: the sign and the approximate amount of the reduction of the $\mathcal{D}$ factor after the changes in axial binding is in full agreement with what previously determined in the case of other Yb complexes (see section 3.6).

6.1.8 Structure and dynamics of the chiral species

The data discussed in the previous sections demonstrate that the chiral complex is a fluxional entity, endowed with a dynamic C_3 symmetry on the NMR timescale. Indeed low temperature experiments (down to -25°C) did not clarify the picture of dynamic processes. We must not forget that fod is a non-symmetric ligand and that the appearence of one resonance for the three t-butyl groups means that either they all point in the same direction (as schematically depicted in Figure 5) or diketonates fastly swap between the two possible orientations. This process further complicates the solution equilibria discussed previously and clarifies that several chiral species are likely to be present. In contrast to NMR, optical spectroscopies are fast observation techniques and thus yield a superposition of signals arising from each form. It is noteworthy that there is a remarkable difference between the two spectral windows, in the UV and in the NIR. The largest contribution to UV CD orginates from the exciton coupling between the electric dipole transition moments of the diketonates, which are in good approximation directed as the line joining the two oxygen atoms.[7] Variations in relevant structural parameters, like the co-ordination number of the lanthanide or the reciprocal positions of t-butyl and perfluoropropyl groups, have a minor influence on the spectra, which depend essentially on the enantiomeric $\Lambda - \Delta$ arrangement of the three fod moieties. NIR CD is sensitive to the crystal field of the lanthanide, consequently it is a reporter of those elements that were overlooked by UV CD, namely number (and nature) of coordinating atoms and to the polarizability of the organic groups (C_3F_7 versus t-Bu) around the metal cation. The spectra in this region are in principle the envelope of up to 12 electronic transition of Yb(III). In most of the present cases they combine to a single line at 975 nm. For diol **13**, we have seen that the observed signal is a superposition of contributions from - at least - two species, characterized by a (negative) peak at 975 nm and a positive Cotton effect at 978 nm. This is the reason for the different responses of the two dichroic techniques: NIR CD highlights the existence of different species and demonstrates the complexity of the problem, UV CD yields homo-geneous spectra for structurally similar compounds, lending itself as a useful configurational correlation. The structural inference may be cautiosly drawn also by NIR CD only when really superimposable spectra are found. For exam-ple, compound **13**, **20** and **21** give very similar NIR CD spectra, dominated by the negative trough at 975 nm and they can be assigned to the same absolute configuration.

6.2 Discussion

The data presented in the previous section demonstrate that $Yb(fod)_3$ both in chloroform (possibly deuterated) and CCl_4 and has a very strong affinity for a wide variety of diols, leading invariably to the formation of a rather stable octa-coordinated chiral complex. Small quantities of water, present as residual in the solvent or even in the hygroscopic lanthanide complex are quantitatively displaced by the diol. This is a most remarkable difference with respect to the original method based on $Yb(dpm)_3$ and is likely to cause the much smoother behaviour of the fluorinated molecule.

In the light of the observed fluxionality of the complex, a consistent hypothesis on the structure of the chiral species can be drawn: the enantiomeric Λ/Δ coordinations of free $Yb(fod)_3$ become diasteromeric in the presence of the chirotopic elements of the diol and an asymmetric transformation of the first kind takes place in solution, which leads to the formation of a complex endowed with well-defined chirality of the ligands.

Unfortunately, the lability of the species prevents the clarification of the origin of the asymmetric induction.

The strong affinity of $Yb(fod)_3$ for diols is such that even very hindered sub-strates can be analysed: thus (R)–1–(9–anthracenyl)–1,2–ethanediol **21** and

(R)–1–(2–naphthalenyl)–1,2–ethanediol **20** give rise to NIR spectra in agreement with their absolute configurations. These diols have a special interest in the literature [38–48], in particular as intermediates in the synthesis of chiral auxilaries for the catalysis of a wide range of asymmetric reactions.

Finally, the diol **19** is an example of a particularly complex molecule that could be analysed by the method proposed. This molecule was synthetised by an asymmetric cis-dihydroxilation from the corresponding olefin according the Sharpless procedure [12] and it constitutes a relevant step in a recent approach in the synthesis of dihydroisocumarins [22].

The sign of the Cotton effect at 976 nm, inverted with respect to all the other R-diols, confirms that this molecule constitutes an exception to the Sharpless memnonic device, which predicts the opposite configuration [22, 34].

The fact that a cyclic molecule as (R, R)–1,2–cyclohexanediol gives a CD spectrum absolutely comparable to that of open-chain systems, ensures that chelation by the two oxygen atoms must occur when the diol is in a g^- conformation, as confirmed also by ^{1}H-NMR analysis. Accordingly, the two bulkiest substituents in the other molecules must be in a pseudo-equatorial position with reference to the 5-membered cycle C–O–Yb–O–C.

6.3   A new method with Yb triflate

Direct observation of the ff transitions of the lanthanide ion enables one to do without chromophoric ligands and to use the most common ionic compounds, such as chlorides and trifluoromethansulfonates (triflates), which are known to give labile adducts with polifunctional species, as aminoacids and sugars [49]. Indeed, by simply adding a four-fold excess of butandiol, **13**, to a 20 mM solution of Yb(TfO)$_3$ in CH$_3$CN, a neat NIR-CD is observed, with a sequence of Cotton effects of alternating signs spanning 80 nm, around 970 nm, as shown in figure 6.9. The spectrum appears immediately after mixing and is indefinitely stable.

One can distinguish 5 main Cotton effects: band I, at around 940 nm (negative); band II, 960 nm (positive, sometimes very weak); band III, 980 nm (strong positive); band IV, 990 nm (strong negative); band V, 1010 nm (weak positive).

With several diols of different structures (see figure 6.10), one obtains the results summarized in Table 6.4.

The correlation between the signs of all the main bands (with the only exception of the weak band II, slightly negative in **23**) of the NIR-CD spectrum and the structure of the diol is remarkable and demonstrates that this method lends itself to investigate the absolute configuration of chiral diols. Also the relative apparent $\Delta\epsilon'$ are rather conserved, although the overall intensities are somewhat varying from a sample to another.

In order to investigate the stoichiometry of the dichroic complex, for **13** full titration was performed. After the first addition of diol (equivalent to a diol-

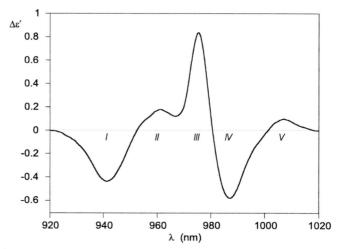

Figure 6.9: NIR-CD of the ytterbium chelate with 2,3-butanediol **13** (Yb/diol=0.25). $\Delta\epsilon'$ is the molar CD referred to the concentration of the diol.

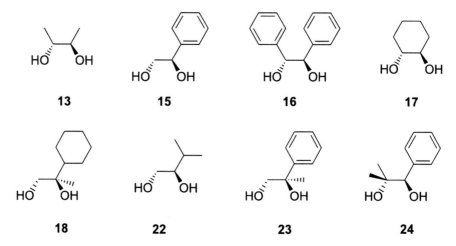

Figure 6.10: Chiral 1,2-diols employed in the analysis with Yb triflate.

to-metal ratio $\rho=0.1$), a weak but detectable CD can be recorded. At low ρ values ($\rho < 1$), the spectrum of Yb-**13** takes the form shown in figure 6.11 and only after an equimolar quantity of diol has been added, it eventually converts to that shown in figure 6.9.

Apparently there are two different chiral chelates in solution whose relative proportion depends on the molar ratio between diol and metal: it seems likely that initially a 1:1 complex is formed, which is then converted into a 2:1. It should be observed, however that the positions and the signs of the most relevant Cotton effects stay the same in the two complexes, which ensures the

diol	Band I	Band II	Band III	Band IV	Band V
13	-4.6	1.0	8.5	-5.8	0.6
15	-0.5	0.0	0.9	-0.7	0.5
16	-3.9	1.6	0.7	-3.6	0.2
17	-13.0	8.2	5.5	-7.5	2.3
18	-1.0	0.4	2.5	-0.8	0.4
22	-7.8	0.0	7.4	-6.1	0.9
23	-1.8	-0.3	1.8	-1.8	0.5
24	-2.8	4.4	2.7	-2.0	0.4

Table 6.4: NIR-CD data of Yb^{3+} chelates with various diols. $\Delta\epsilon'$ (scaled for the concentration of diol) are reported for the 5 main Cotton effects, indicated in the text and in figure 6.9.

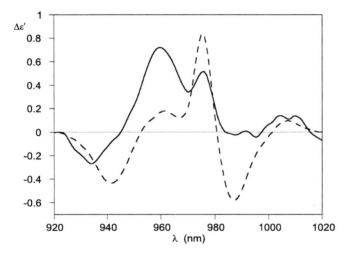

Figure 6.11: Same as in figure 6.9, but with a ratio Yb/diol=2.

safe assignment of absolute configuration, whatever the relative proportion of the two species.

The nature of the the chiral complexes deserves a deeper investigation. ^{19}F-NMR demonstrates that the triflate ion is partly bound to the paramagnetic centre, suggesting that it partecipates in completing the coordination sphere of the lanthanide ion.

A strong chelate effect must be present: no NIR-CD is induced upon addition of a large excess of (S)-2-butanol; moreover, exactly the same spectrum as that reported in figure 6.9 can be observed even when (R,R)-butandiol is added to an ethanolic solution of $Yb(TfO)_3$.

6.4   Conclusions

It was demonstrated that the use of the fluorinated ytterbium compound $Yb(fod)_3$ offers a useful method for the determination of the absolute configuration of chiral diols. The method is very simple and straightforward, since no derivatization is necessary, but a simple 1:1 mixing of analyte and auxiliary.

The use of ytterbium opens the way to NIR CD investigation, which is particularly convenient since it sheds light on the fluxionality and complexity of the system. Moreover, in case of close spectral analogy, it can be cautiosly taken as a stereochemical inference. The manifold of molecules investigated in the present work adds up to those previously reported in the literature, demonstrating pliancy and broad scope for the UV CD correlation. A general, non-empirical method for open-chain diols is not available and probably will never be found, because structural variations in the nearby groups introduce elements of novelty and originality in the molecules, which will force one to resort to complex analyses. Empirical approaches as the present one, on the other hand, may look very attractive because of their extreme ease of operation and sensitivity. Naturally, as pointed out by Harada and Nakanishi, empirical correlation must be used with extreme care: one way to circumvent the problem is to have several such methods at hand: if no single correlation can be regarded as ultimate, the simultaneous use of several may give a reasonable confidence and the demonstration of the scope of a new structural inference may thus be welcome.

Two spectral windows, in the UV and in the NIR, ensure pliancy and safety of the technique: in the first place, strongly absorbing systems can be analysed; secondly, the two-fold correlation (negative band at 300 and 976 nm) $\leftrightarrow$ (bR) or (bR, bR) [5] diol configuration ensures an internal consistency check. We have not found exceptions to the UV CD correlation, which on ambiguous cases should thus prevail over the NIR.

The method of Yb triflate allows the determination of the absolute configuration of aliphatic and aromatic diols and offers a useful complement and alternative to other methods. It is fast and very straightforward, since it requires simple mixing of the substrate with the inexpensive $Yb(TfO)_3$.

In addition, the diol can be quantitatively recovered by a simple extraction in ether after washing the solution with water. The resulting CD spectrum is stable over a long time, which allows one to record multiple scans to improve the signal-to-noise ratio.

When analyzing a new diol by this method, the presence of several diagnostic bands and the sensitivity of the NIR-CD spectrum of Yb(III) to the chemical nature of the ligand, enables one to recognize the formation of a chelate analogue to those presented here. No exception to the sequence of signs reported on Table 6.4 was found, which ensures the fast and reliable assignment

---

[5]For an explanation of this generalized stereochemical notation, see ref. [23].

of absolute configuration of unknown diols.

A general, non-empirical method for open-chain diols is not available and probably will never be found, because structural variations in the nearby groups introduce elements of novelty and originality in many molecules, which will force one to resort to complex analyses. Empirical approaches as the present ones, on the other hand, may look very attractive because of their extreme ease of operation and sensitivity. Naturally, as pointed out by Harada and Nakanishi [50], empirical correlation must be used with extreme care: one way to circumvent the problem is to have several such methods at hand: if no single correlation can be regarded as ultimate, the simultaneous use of several may give a reasonable confidence and the demonstration of the scope of a new structural inference may thus be welcomed.

Bibliography

[1] NAKANISHI, K.; SCHOOLEY, D. A.; KOREEDA, M.; DILLON, J., *J. Chem. Soc., Chem. Commun.* **1971**, 1235–1236.

[2] NAKANISHI, K.; DILLON, J., *J. Am. Chem. Soc.* **1971**, *93*, 4058–4060.

[3] BUKHARI, S. T. K.; GUTHRIE, R. D.; SCOTT, A. I.; WRIXON, A. D., *J. Chem. Soc., Chem. Commun.* **1968**, 1580–1582.

[4] BUKHARI, S. T. K.; GUTHRIE, R. D.; SCOTT, A. I.; WRIXON, A. D., *Tetrahedron* **1970**, *26*, 3653–3659.

[5] NELSON, W. L.; WENNERSTRÖM, J. E.; SANKAR, S. R., *J. Org. Chem.* **1977**, *42*, 1006–1012.

[6] SCOTT, A. I.; WRIXON, A. D., *J. Chem. Soc., Chem. Commun.* **1969**, 1184–1186.

[7] SNATZKE, G.; WAGNER, U.; WOLFF, H. P., *Tetrahedron* **1981**, *37*, 349–361.

[8] FRELEK, J.; GEIGER, M.; VOELTER, W., *Curr. Org. Chem.* **1999**, *3*, 117–146 AND REFERENCES THEREIN.

[9] DILLON, J.; NAKANISHI, K., *J. Am. Chem. Soc.* **1974**, *96*, 4057–4059.

[10] DILLON, J.; NAKANISHI, K., *J. Am. Chem. Soc.* **1975**, *97*, 5409–5417.

[11] DILLON, J.; NAKANISHI, K., *J. Am. Chem. Soc.* **1975**, *97*, 5417–5422.

[12] KOLB, H. C.; VAN NIEUWEHNZE, M. S.; SHARPLESS, K. B., *Chem. Rev.* **1994**, *94*, 2483–2547.

[13] HALE, K. J.; MANAVIAZAR, S.; PEAK, S. A., *Tetrahedron Lett.* **1994**, *35*, 425–428.

[14] CARREIRA, E. M.; BOIS, J. D., *J. Am. Chem. Soc.* **1994**, *116*, 10825–10826.

[15] KRYSAN, D. J.; ROCKWAY, T. W.; HAIGHT, A. R., *Tetrahedron: Asymm.* **1994**, *5*, 625–632.

[16] IWASHIMA, M.; KINSHO, T.; III, A. B. S., *Tetrahedron Lett.* **1995**, *6*, 2199–2202.

[17] KRYSAN, D. J., *Tetrahedron Lett.* **1996**, *37*, 1375–1376.

[18] VANHESSCHE, K. P. M.; SHARPLESS, K. B., *J. Org. Chem.* **1996**, *61*, 7978–7979.

[19] BOGER, D. L.; MCKIE, J. A.; NISHI, T.; OGIKU, T., *J. Am. Chem. Soc.* **1996**, *118*, 2301–2302.

[20] BOGER, D. L.; MCKIE, J. A.; NISHI, T.; OGIKU, T., *J. Am. Chem. Soc.* **1997**, *119*, 311-325.

[21] ERTEL, N. H.; DAYAL, B.; RAO, K.; SALEN, G., *Lipids* **1999**, *34*, 395–405.

[22] SALVADORI, P.; SUPERCHI, S.; MINUTOLO, F., *J. Org. Chem.* **1996**, *61*, 4190–4191.

[23] ELIEL, E. L.; WILEN, S. H.; MANDER, L. N., *Stereochemistry of Organic Compounds;* WILEY: NEW YORK, 1994.

[24] DAYAL, B.; SALEN, G.; PADIA, J.; SHEFER, S.; TINT, G. S.; WILLIAMS, T. H.; TOOM, V.; SASSO, G., *Chemistry and Physics of Lipids* **1992**, *61*, 271–281.

[25] DAYAL, B.; KESHAVA, R.; SALEN, G.; SEONH, W. M.; PRAMANIK, B. N.; HAUNG, C. E.; TOOME, V., *Pure & Appl. Chem.* **1994**, *66*, 2037–2040.

[26] PARTRIDGE, J. J.; TOOME, V.; USKOKOVIČ, M. R., *J. Am. Chem. Soc.* **1976**, *98*, 3740–3741.

[27] PARTRIDGE, J. J.; SHIUEY, S.; CHADHA, N. K.; BAGGIOLINI, E. G.; BLOUNT, J. F.; USKOKOVIČ, M. R., *J. Am. Chem. Soc.* **1981**, *103*, 1253–1255.

[28] DAYAL, B.; TINT, G. S.; SHEFER, S.; SALEN, G., *Steroids* **1979**, *33*, 327–338.

[29] DAYAL, B.; SALEN, G.; TINT, G. S.; TOOME, V.; SHEFER, S.; MOSHBACH, E. H., *J. Lipids Research* **1978**, *19*, 187–190.

[30] DAYAL, B.; SALEN, G.; TOOME, V.; TINT, G. S., *J. Lipids Research* **1986**, *27*, 1328–1332.

[31] TSUKUBE, H.; HOSOKUBO, M.; WADA, M.; SHINODA, S.; TAMIAKI, H., *J. Chem. Soc., Dalton Trans.* **1999**, 11–12.

[32] TSUKUBE, H.; SHINODA, S., *Enantiomer* **2000**, *5*, 13–22.

[33] TSUKUBE, H.; HOSOKUBO, M.; WADA, M.; SHINODA, S.; TAMIAKI, H., *Inorg. Chem.* **2001**, *40*, 740–745.

[34] DI BARI, L.; PESCITELLI, G.; PRATELLI, C.; SALVADORI, P., *J.*

Org. Chem. **2001**, *66*, 4819–4825.

[35] SCOTT, A. I.; WRIXON, A. D., *Chem. Commun.* **1969**, 1184–1186.

[36] ABRAHAM, R. J.; CASTELLAZZI, I.; SANCASSAN, F.; SMITH, T. A. D., *J. Chem. Soc., Perkin Trans. 2* **1999**, 99–106 AND REFERENCES THEREIN.

[37] BATSANOV, A. S.; BEEBY, A.; BRUCE, J. I.; HOWARD, J. A. K.; KENWRIGHT, A. M.; PARKER, D., *Chem. Comm.* **1999**, 1011–1012.

[38] BECKER, H.; KING, S. B.; TANIGUCHI, M.; VANHESSCHE, K. P. M.; SHARPLESS, K. B., *J. Org. Chem.* **1995**, *60*, 3940–3941.

[39] COREY, E. J.; NOE, M. C.; GROGAN, M. J., *Tetrahedron Lett.* **1994**, *35*, 6427–6430.

[40] COREY, E. J.; NOE, M. C.; GUZMAN-PEREZ, A., *J. Am. Chem. Soc.* **1995**, *117*, 10817–10824.

[41] COREY, E. J.; NOE, M. C., *J. Am. Chem. Soc.* **1996**, *118*, 11038–11053.

[42] NICOLAU, K. C.; RENAUD, J.; NANTERMET, P. G.; COULADOUROS, E. A.; GUY, R. K.; WRASIDLO, W., *J. Am. Chem. Soc.* **1995**, *117*, 2409–2420.

[43] TORII, S.; LIU, P.; BHUVANESWARI, N.; AMATORE, C.; JUTLAND, A., *J. Org. Chem.* **1996**, *61*, 3055–3060.

[44] CHO, B. T.; CHUN, Y. S., *J. Org. Chem.* **1998**, *63*, 5280–5282.

[45] MIAO, G.; ROSSITER, B. E., *J. Org. Chem.* **1995**, *60*, 8424–8427.

[46] KAWASAKI, K.; KATSUKI, T., *Tetrahedron* **1997**, *53*, 6337–6350.

[47] WANG, G.; PFALTZ, T.; MINDER, B.; MALLAT, T.; BAIKER, A., *J. Chem. Soc., Chem. Commun.* **1994**, 2047–2048.

[48] SCHÜRCH, M.; HEINZ, T.; AESHIMANN, R.; MALLAT, T.; PFALTZ, A.; BAIKER, A., *Tetrahedron: Asymm.* **1990**, *1*, 221.

[49] BRITTAIN, H. G., *Coord. Chem. Rev.* **1983**, *48*, 243–276.

[50] HARADA, N.; NAKANISHI, K., *Circular Dichroic Spectroscopy – Exciton Coupling in Organic Stereochemistry*; OXFORD UNIVERSITY PRESS: OXFORD, 1983 (CHAPTER 9).

Chapter 7

Lanthanides as substitutes for "silent" metal ions

7.1 Yb^{3+} complexes with anthracycline antitumor antibiotics

The anthracycline antibiotics isolated from several *Streptomyces* species are widely used as antibiotics for medical treatment of human tumors [1–4]. Daunorubicin (Dau, see figure 7.1) is the prototypical member in this family of drugs, all of which consist of a 4-ring anthraquinone chromophore and an amino sugar substituent at the 7-position [4].

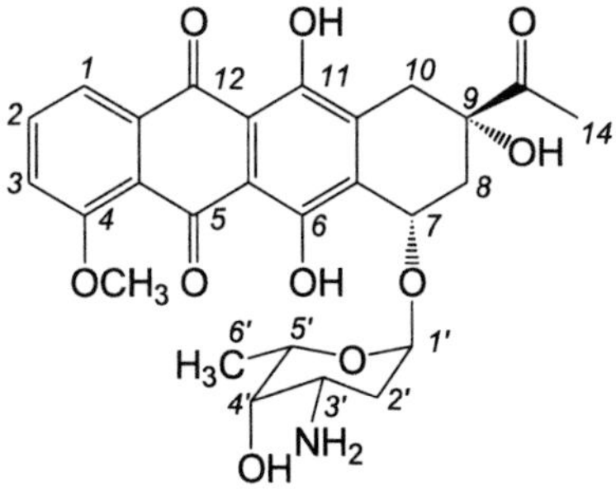

Daunorubicin (**25**)

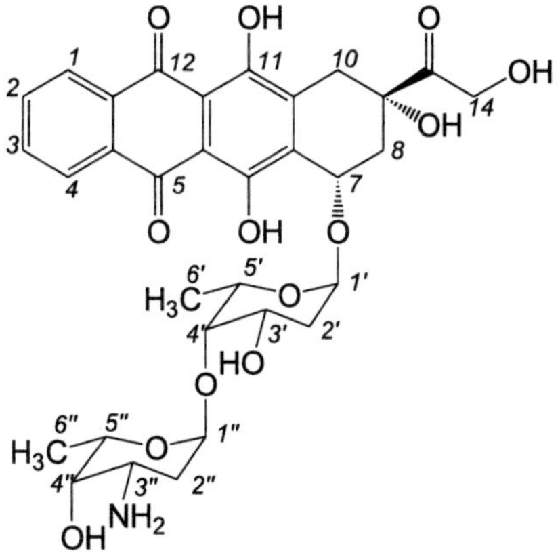

MEN 10755 (**26**)

Figure 7.1: Structure and atom numeration of daunorubicin (**25**) and MEN 10755 (**26**).

Their bioactivity is due essentially to the ability to bind DNA and to the redox activity of the anthraquinone group [5, 6] and occurs in two steps, through:

- an intercalation of the drugs into the base pairs in the DNA minor groove;

- a free radical damage of the ribose, where the free radicals are formed during the redox cycle of the anthraquinone.

They are also known to form complexes with various metals and notably the iron adduct has been held responsable for the cardiotoxicity of anthracyclines [7, 8]. Reducing the occurrence and the severity of these side effects is a central objective of current research on these clinical remedies [7–12].

It is clear that, in order to improve the knowledge on the structural origin of the clinical properties of anthracyclines, a conventional conformational investigation is largely insufficient and must be completed with a study of the interaction with different cations[7, 10, 13–21].

In this context there has been a prominent interest toward trivalent lanthanide ions (Ln^{3+}), owing to their unique spectroscopic properties. In spite of their limited direct involvement in biological system, Ln^{3+} lend themselves as powerful structural probes to highlight the features of cation-binding sites on the organic molecules. In fact, it is often claimed that lanthanides may effectively simulate biologically relevant metal ion systems, as Fe^{3+}, Mn^{2+}, Mg^{2+} or frequently Ca^{2+} [22].

The advantages achieved in structural investigations by sustituting other metal cations with lanthanide ions are connected tightly to electronic features (intraconfigurational transitions in 4f orbitals, paramagnetism, fluorescence) and to greater complex formation constants compared to Ca^{2+} [22, 23], which leads to increased concentration of the bound form. Currently, several techniques accessible for chemical, biochemical or biological studies involving these ions are described [22, 24, 25]. By UV-Vis absorption and circular dichroism or NMR spectroscopies, for example, it is possible to study the modifications of the organic moieties following to the lanthanide ions coordination. Particularly, most trivalent lanthanide ions (La^{3+} and Lu^{3+} excepted) are paramagnetic. This property is widely exploited in NMR structural investigation, because of two paramagnetic effects (lanthanide induced shifts, LIS, and lanthanide induced relaxation, LIR), directly connected with geometric parameters [24, 26]. On other hand, the binding of Ln^{3+} with an organic molecule influences the intraconfigurational 4f transintions of the ion itself, whereupon, in a chiral enviroment (and in presence of a stereodefined complex), Ln^{3+} can exhibit intense induced circular dichroism or circularly polarized luminescence bands [27].

The rich work performed by Rueben and Lenkinski [28] on adriamycin and successively by Ming [29, 30] on daunomycin answered some questions regarding the metal binding and the structure of the complexes and shed light on the identity of different species detected in the spectra, the stoichiometry of the metal-drug complexes and the metal binding mode.

In particular, a comprehensive report has recently been published on Yb^{3+} binding with daunomycin: optical and NMR experiments demonstrate that

Yb^{3+} is primarily bound to the 11,12-β-ketophenolate site with a conformation similar to the free drug and, depending upon the proton activities, several different metal–drug complexes with 1:1, 1:2, 1:3 and 2:1 metal–to–drug ratios are formed in both acquous and methanol solutions.

Although a large amount of data has been collected on the adducts formed in solution, no complete molecular simulation has ever been attempted on the basis of such data, that is:

- the measured relaxation times of the protons in the different paramagnetic species were only employed for a *qualitative* location of the metal ion; moreover, the Curie contribution to relaxation was inexplicably neglected (a noticeable wrong extimate in ref. [28]!), so that the analysis was performed only on the basis of the pseudocontact mechanism;

- the values and orientation of the magnetic susceptibility anysotropy tensor has never been determined.

The success of this kind of investigation centers on the assumption that replacing of Ca^{2+} by Ln^{3+} is really isostructural. Such a requirement, although constitutes the necessary premise before any susequent study of Ca^{2+}-biomolecule interaction through lanthanide probes, is rarely verified experimentally. Here follows an account on a comparative spectroscopic study of interaction with metal ions of MEN 10755 (**26**, figure 7.1), a new antitumor disaccharide analogue of doxorubicin currently object of pharmacologic and clinical investigations [31–33], which lends itself as potential cation binding biomolecule. Specifically, the paramagnetic Yb^{3+} ion was used as informative probe for ^{1}H-NMR, UV-Vis absorption and UV-Vis or NIR CD, after rapidly verifying by spectroscopies the isomorphism between Ln^{3+} complexes (Yb^{3+}, La^{3+}, Lu^{3+}) and Ca^{2+} complex.

The choice of ytterbium(III) as the principal probe for the structural study is justified by its peculiar spectroscopic properties, as widely discussed in section 2.3 and 1.3.

7.2 Dimerization constant and structure of 26 in solution

In different reports, it is shown that anthracyclines tend to self-associate in aqueous solution, but not in methanol or in solvents mixtures water/ethanol [34–37].

In agreement to other antracyclines we found that **26** does not self-associate in methanol but dimerizes in water. We estimated the dimerization constant at pH 7.6 by optical tritration, as well.

In figure 7.2 is reported the apparent molar absorptivity ($\epsilon' = \frac{A}{c_0}$) versus the analitical concentration of **26** (c_0) in aqueous pH 7.6 buffered solution. Equation 7.1 [34] describes the dependence of (ϵ') on (c_0):

$$\epsilon' = \epsilon_D + \frac{\epsilon_M - \epsilon_D}{4K_{dim}c_0}(\sqrt{1 + 8K_{dim}c_0} - 1) \tag{7.1}$$

where K_{dim} is the dimerization costant, ϵ_D and ϵ_M are the molar absorptivities of dimer and monomer, respectively. By fitting the experimental data with equation 7.1, we evaluated simultaneously K_{dim} , ϵ_D and ϵ_M as fitting parameters. The values of these parameters, reported in figure 7.2, are consistent with those of other anthracyclines [34–37].

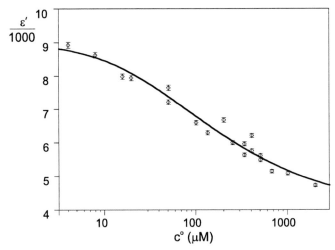

Figure 7.2: Apparent molar absorptivity (ϵ') at 485 nm versus analitical concentration of **26** (c_0) in aqueous pH 7.6 TRIS buffer solution. The solid line represents the best fit in agreement with the equation 7.1, reported in the text. In the table, the fitting parameter ϵ_D, ϵ_M (molar absorptivities of dimer and monomer, respectively) and K_d (dimerization costant) are indicated. The spectra were recorded using optical path of 0.05 cm or 0.10 cm, for concentrated solutions (> 800 mM); 0.20 cm and 0.50 cm, for intermediate concentrations (400-700 mM); 1.00 cm and 10.00 cm, for dilute (50-200 mM) and very dilute ($\simeq$20 mM) solutions, respectively.

The ^{1}H-NMR spectrum of **26** in D_2O is also strongly concentration dependent (table 7.1), whereby the limit for infinite dilution very closely resembles the proton spectrum in methanol. Therefore, the molecule in the latter solvent can be regarded as a good model for the monomeric species[1]. Standard 2D-NMR techniques (COSY and NOESY) followed by MM2 structural optimization with NOE constraints yields the conformation depicted in figure 7.7.

The involvement of the aromatic portion in the dimerization process is demonstrated by the stronger variation of the resonances of these proton in concentrated solution, as shown in table 7.1. In fact, in a concentrated D_2O solution

---

[1]We found similarities also between 1H-NMR spectra of Yb·**26** in CD_3OD and in D_2O.

the signals of the all aromatic protons ($C_{1-4}H$) are pratically isochronous (7.60 ppm). Dilution of the same solution leads to differentiation of signal to 8.14 ppm and 7.81 ppm; likewise in methanol one finds 8.16 ppm and 7.82 ppm. These results indicate clearly that for this new anthracycline the stacking of molecules is also the most reasonable dimerization mechanism.

7.3   Stoichiometry of Yb·26 complexes

The changes of the absorption spectrum of compound **26** in presence of Yb^{3+} ion are indicative of the formation of Yb·**26** complex both in aqueous solution and in methanol. In fact, addition of Yb^{3+} induces a shift in λ_{max} from 486 nm to 536 nm in a pH 7.6 aqueous solution, associated to an immediate change of color from orange to purple. In the case of a methanolic solution, after addition of Yb^{3+} a shift of absorption maximum from 482 nm to 519 nm is observed, with the appearance of a resolved fine structure and the change of color from orange to red.

In order to determine the stoichiometry ratio of the Yb·**26** complex, we realized spectrofotometric titrations in buffered (pH 7.6) aqueous or methanolic solutions of **26** with Yb^{3+}. The analysis of the experimental absorption data by the molar ratio method indicates the formation of 1:3 and 1:1 Yb·**26** complexes in water (figure 7.3(a)). In methanol it is evident the predominant formation of 1:1 Yb·**26** complex and eventually, for higher metal/drug ratio, a 3:1 different complex (figure 7.3 (b)).

Following equations 7.2 and 7.3 for molar ratios below 1.5, the analysis of absorption at 571 nm, where the absorption of the free drug is neglegible, led to estimate the complex formation constant for Yb·**26**:

$$\frac{A}{l} = \epsilon_C \left[Yb \cdot \mathbf{26} \right] \tag{7.2}$$

$$[Yb \cdot \mathbf{26}] = \frac{(K_c c_0 (n+1) + 1) - \sqrt{(K_c c_0 (n+1) + 1)^2 - 4(K_c c_0)^2 n}}{2K_c} \tag{7.3}$$

where A is the absorptivity; l is the optical path; ϵ_C is the molar absorptivity of complex, whose concentration is indicated between square brackets; c_0 is the total concentration of **26** and n corresponds to the Yb·**26** ratio, that is variable during titration.

Before further characterizing the potential calcium-binding drug, **26**, by using Yb^{3+} probe, isomorphism between complexes of **26** with Ca^{2+} and Ln^{3+} was checked through a spectroscopic investigation, as reported below.

First of all, information about isomorphism of such complexes were obtained by comparing the Vis absorption and CD spectra of methanolic solutions of M^{n+}·**26** 1:1 ($M^{n+} = Ca^{2+}$, La^{3+}, Lu^{3+}, Yb^{3+}) with the corresponding spectra of free ligand (figure 7.4).

From experimental absorption spectrum of Yb·**26** 1:1 in methanol the contribution of the free drug (67%, calculated by the previously estimated K_c) was

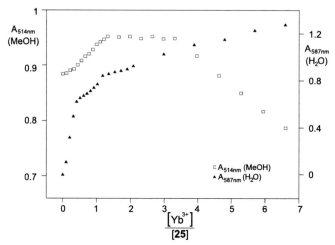

Figure 7.3: Molar ratio plot relative to the titration of **26** with $YbCl_3$ in aqueous (pH 7.6, triangles) and in methanolic (squares) 50 mM TRIS buffer solutions. A^{587nm} and A^{514nm} are the experimental absorptivity at 587 nm and at 514 nm, where that of free drug is negligible. In the two cases, pairs of isosbestic points were found at $\simeq$496 nm and $\simeq$536 nm and at 504 nm and 514 nm. ($YbCl_3$ was added in subsequent aliquotes of 5.6 ml, 10.0 ml and 50.0 ml of a 7350 mM solution into 2.0 ml of **26** 203 mM in the first case; as aliquotes of 15 ml, 45 ml and 90 ml of a 2193 mM solution into 2.0 ml of **26** 155 mM, in the second case; optical path 1 cm)

subtracted. In this way, the absorption spectrum of this complex was isolated (figure 7.5).

A linear combination of the spectra of Ca·**26** 1:1 and free **26** in methanol, with relative weights 1 and 0.98 respectively, led to the calcium complex spectrum, in which it is possible to recognize a close likeness to that of Yb·**26**. This result allows one to recognize identity of spectra of both complexes and to estimate an amount less than 2% of Ca·**26** 1:1, in agreement to what reported for daunorubicin [17].

By the same procedure, it was calculated that around 10% of **26** forms the corresponding complex with La^{3+} or Lu^{3+}.

Once evaluated the mole fractions of the complexes in methanolic solution, we roughly estimated their molar absorptivities (ϵ) for the maximum absorption wavelenght in agreement with

$$\epsilon_{M/1} = \frac{\epsilon' - \epsilon_1(1-x)}{x} \qquad (7.4)$$

where ϵ' is the apparent molar absorptivity, $\epsilon_{M/1}$ is the molar absorptivity of the complex and ϵ_1 that of free **26**.

The calculated values are rather similar (allowing for the large error arising from the uncertainty in the fraction of the bound form): $\epsilon_{Ca·26} \simeq 47000$

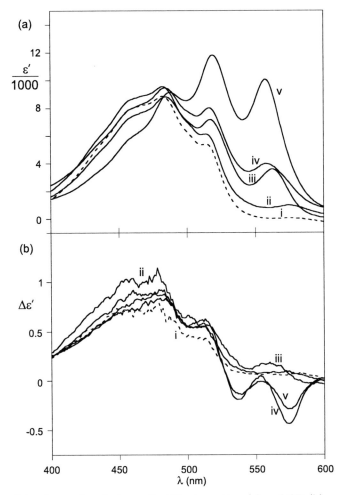

Figure 7.4: Comparison between the Vis absorption (a) and CD (b) spectra of **26** and those of $M^{n+} \cdot$**26** 1:1 systems in methanol, with concentrations of about 300 mM. (i) **26** (dashed line); (ii) Ca·**26**; (iii) La·**26**; (iv) Lu·**26**; (v) Yb·**26**; optical path 1 cm.

$M^{-1}\ cm^{-1}$; $\epsilon_{La \cdot 26} \simeq 29000\,M^{-1}cm^{-1}$; $\epsilon_{Lu \cdot 26} \simeq 49000\,M^{-1}cm^{-1}$; $\epsilon_{Yb \cdot 26} \simeq 30000\,M^{-1}cm^{-1}$.

These observations underline that the spectral changes of metanolic solutions of **26** depend on the coordination of the cation, but not on the kind of ion. Furthermore, it is evident that the lower value of K_c for Ca^{2+} is the cause of the weak spectroscopic effects of coordination in this case.

The isomorphism between the four 1:1 metal complexes of **26** was investigated more in detail by ^{1}H-NMR of their d_4-methanolic solutions. Comparing the ^{1}H-NMR spectra of these complexes, one immediately observes two different

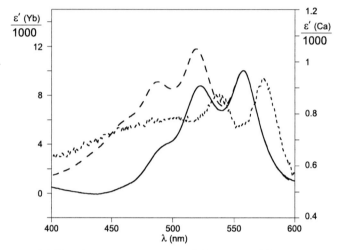

Figure 7.5: (a) Vis absorption spectrum of the system Yb·**26** 1:1 in methanol (dashed line) and that of the complex (33% of bound **26**), calculated as indicated in the text (solid line). (b) Vis absorption spectrum of the Ca·**26** 1:1 complex in methanol (dotted line), calculated as explained in the text and corresponding to 2% of bound drug.

apparent exchange kinetics. On the NMR time scale, there is slow exchange for the paramagnetic Yb^{3+} complex ($\nu_{exchange} \ll 50s^{-1}$), while it is fast for the diamagnetic ions (Ca^{2+}, La^{3+}, Lu^{3+}). This difference is due to the larger range of shifts induced by the unpaired electron and not a real variation of kinetic behavior. Indeed, the proton NMR spectrum of Yb·**26** 1:1 shows one set of well resolved signals in the diamagnetic region (from 1.19 to 8.34 ppm), assigned to the free form of the drug (see table 7.1), and one set of seventeen signals in the high field region (from 0.88 ppm to -38.06 ppm), assigned to the protons of the complexed form of **26**.

Comparing the integrals of free and Yb^{3+}-bound (paramagnetically shifted) forms, one obtains a ratio approximately 3:1, in good agreement with the UV data. By inspection of the 0-10 ppm part of the ^{1}H-NMR spectra of **26** free and after addition of Ca^{2+} and Ln^{3+} we can observe that (table 7.1):

- some of the signals of the free drug are shifted by cations addition (notably, H-1, H-4, H-7 H-10 on the aglicone moiety);

- addition of different diamagnetic cations induce the *same* spectral changes;

- Yb^{3+} binding differentiates between a truly metal bound form (whose resonances are paramagnetically shifted) and a "free" diamagnetic molecule, whose spectrum is different from that of isolated **26** and practically identical to the one of the Lu^{3+} adduct.

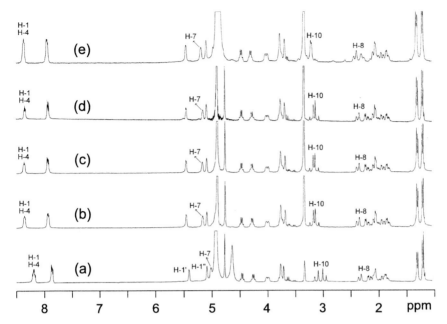

Figure 7.6: ^{1}H-NMR spectrum in CD$_3$OD of: (a) **26**; (b) Ca·**26** 1:1; (c) La·**26** 1:1; (d) Lu·**26** 1:1; (e) Yb·**26** 1:1. Labels of complexation diagnostic peaks are reported.

We conclude that, upon metal cation addition two different processes take place. First, a fast deprotonation of a hydroquinone group on **26** (indicated in the following equation as LH) to yield its anionic form occurs:

$$LH \rightleftharpoons L^- + H^+$$

This takes place in a buffered solution and is promoted by the following step, reversible cation binding:

$$L^- + M^{n+} \rightleftharpoons LM^{(n-1)+}$$

The latter process has two distinct kinetic regimes with respect to the NMR timescale: it is *fast* for diamagnetic ions, *slow* for Yb^{3+}. Such a difference is only apparent and does not imply different rates, because it is due to the very large magnitude of the shifts induced by the unpaired electron. The fact that a distinct paramagnetically shifted signal is detected for H$_{ax}$-C$_{2'}$ in the Yb^{3+} adduct, with a pseudocontact term of -3.50 ppm, sets an upper limit to the exchange rate to about 1050 s^{-1}.

	δ (ppm)						
	26			$M^{n+}\cdot$**26** 1:1 in CD_3OD			
^{1}H	D_2O		CD_3OD	Ca^{2+}	La^{3+}	Lu^{3+}	Yb^{3+}
	conc	dil					
C_1H, C_4H	7.6	8.14	8.16	8.30	8.33	8.30	8.34
C_2H, C_3H	7.6	7.81	7.82	7.90	7.91	7.90	7.92
$C_{1'}H$	5.26	5.40	5.37	5.43	5.42	5.43	5.42
$C_{1''}H$	5.03		5.05	5.05	5.05	5.05	5.06
C_7H		4.97	5.13	5.12	5.10	5.15	
$C_{14}H_2$		4.73	4.71	4.71	4.73	4.79	
$C_{5''}H$	4.35	4.46	4.41	4.43	4.42	4.43	4.45
$C_{5'}H$	4.08	4.26	4.22	4.24	4.23	4.23	4.26
$C_{3'}H$	3.93		3.96	3.97	3.96	3.95	3.98
$C_{3''}H$	3.76	3.69	3.70	3.74	3.70	3.72	3.74
$C_{4'}H$	3.67		3.70	3.73	3.72	3.72	3.74
$C_{4''}H$	3.72	3.65	3.65	3.65	3.64	3.66	3.66
$C_{10}H_2$	2.76		3.07	3.16	3.17	3.18	3.19
	2.47		2.93	3.07	3.06	3.04	
C_8H_2	2.15	2.15-1.70	2.30	2.35	2.32	2.32	2.39
	1.90		2.12	2.18	2.21	2.09	2.28
$C_{2''}H_2$	1.95	2.15-1.70	2.05	2.06	2.02	2.03	2.05
$C_{2'}H_2$	1.80	2.15-1.70	1.86	1.87	1.88	1.86	1.86
$C_{6'}H_3$	1.20	1.17	1.27	1.27	1.27	1.26	1.29
$C_{6''}H_3$	1.06	1.05	1.17	1.18	1.17	1.18	1.19

Table 7.1: ^{1}H shifts (δ, ppm, referred to TMS) of **26** both in D_2O (concentrated and diluted solution) and in CD_3OD, and of $M^{n+}\cdot$**26** 1:1 systems in CD_3OD (in the case of Yb^{3+} only diamagnetic shifts are reported).

7.4   Metal ion binding site of 26

In the case of Yb·**26** 1:1, it was not possible to correlate the signals of the metal-bound with those of the free form of the drug by bidimensional magnetization transfer technique (EXSY), as proposed in similar cases by Ming[18], because of ineffective transfer due to slow kinetics. Moreover a slow formation of a second complex during 2D-acquisition complicated further the spectrum.

However, for the seven more shifted signals of the bound form we found a very good agreement with the ^{1}H–NMR data published in the case of Yb·**25** 1:1 complex [18] and assigned by EXSY. It is reasonable to assume that the paramagnetic ion is coordinated in similar manner in the two systems; we can thus exploit the larger amount of experimental informations available on the latter molecule for a rigorous numerical analysis.

In table 7.3, the experimental values of T_1 and δ_{exp}^{pc} for Yb·**25** 1:1 in methanol

Yb·26 1:1 (bound form)	
-38.1	$C_{10}H$
-32.1	$C_{10}H$
-9.3	C_8H
-6.8	C_7H
-6.6	C_8H
-4.8	C_1H
-3.8	C_2H_2
-2.0	
-2.0	
-1.9	
-1.6	
-0.9	
-0.4	
-0.3	
0.4	
0.9	

Table 7.2: ^{1}H shifts (δ, ppm, referred to TMS) of the bound form of Yb·**26** 1:1 in CD_3OD (known assignment are reported between round brackets).

are reported, together with the corresponding quantities (T_1') corrected for the diamagnetic reference (average $T_1^{dia} = 0.5$ s)^2.

By means of the program PERSEUS (see appendix A.6.1) employing the relaxation and shift data for Yb·**25** the metal ion was positioned on a site corresponding to a coordination by the O-11, O-12 quinone system (figure 7.7), in agreement with what previously hypotised [28]. The set of proton-Yb^{3+} distances obtained is also reported in table 7.3, as well as those estimated by T_1' with the fitted generalized relaxation constant (see sections 2.2.3 and A.6.1) $C = (2.0 \pm 0.2) \cdot 10^{-13}$ s, where the error is indicated as standard deviation. Furthermore, the optimization algorithm allowed us to evaluate the components of the magnetic susceptibility anisotropy tensor from the experimental pseudocontact shifts. The calculated shifts values are reported in table 7.3as well, and feature a good average agreement factor ($\langle|\delta^{pc}_{exp} - \delta^{pc}_{calc}|\rangle = 0.8$ ppm); the fitted values of the $\tilde{\chi}$ components are reported in table 7.4, both in the molecular reference system and in the principal axes system of the tensor. These results show that in this complex the anthracycline protons are at least five bonds remote from Yb, which is consistent with the approssimation to neglect the contact contribution to the shift. In the case of Yb·**26** the coordination geometry through PERSEUS is prevented owing to the limited number

---

^{2}Lu^{3+}·**26** 1:1 resonances were chosen as diamagnetic references instead of those of compound **26**, to take into account differences induced by complexation.

^{1}H	δ^{pc}_{exp} (ppm)	δ^{pc}_{calc} (ppm)	T_1 (ms)	T'_1 (ms)	r (Å)	r_{calc} (Å)
C_1H	-12.3	-11.8	12	12	4.4	4.4
C_7H	-11.5	-9.4	217	383	7.9	7.6
C_8H_2	-10.8	-11.9	150	214	7.1	6.8
	-8.7	-8.4	180	281	7.5	8.2
$C_{10}H_2$	-39.2	-39.7	16	16.5	4.6	4.7
	-33.8	-33.1	20	21.7	4.9	4.9
$C_{1'}H$	-7.2	-5.9	232	433	8.0	9.3
$C_{5'}H$	-5.0	-6.6	263	555	8.4	9.3
$C_{6'}H_3$	-2.8	-3.3	357	1250	9.6	11.4
$C_{14}H_3$	-4.5	-5.0	183	289	7.5	8.6
$C_{2'}H_2$	-5.7	-4.6				10.1
	-3.9	-3.6				11.3
$C_{3'}H$	-4.3	-6.2				9.6
OCH_3	-1.4	-2.1				9.3
$C_{4'}H$	-2.7	-3.9				11.1
C_3H	-3.0	-2.7				8.2
C_2H	-2.6	-3.7				6.8

Table 7.3: ^{1}H-NMR data of Yb·**25** 1:1 in CD_3OD derived from experimental values (δ_{pc}, T_1, T'_1 and r) and calculated pseudocontact shifts (δ^{calc}_{pc}) and proton-ion distances (r_{calc}), as indicated in the main text. values of the δ^{pc}_{exp} are derived from ref. [18], the relaxation times T_1 are taken from ref. [28].

of LIR and LIS experimental values. However, the location of Yb^{3+} must be closely similar to Yb·**25**, as witnessed by the similarity of the pseudocontact shifts (tables 7.3 and 7.5).

$\tilde{\chi}$ components			
molecular reference system		principal axes system	Euler angles
$\chi_{zz} - \frac{\chi_{xx}-\chi_{yy}}{2}$ -1345 ± 150	χ_{xx} -400 ± 50	α 105 ±15 °	
$\frac{3}{2}(\chi_{xx} - \chi_{yy})$ -810 ± 10	χ_{yy} -1870 ± 180	β 45 ± 5 °	
χ_{xy} 1310 ± 100	χ_{zz} 2270 ± 200	γ 33 ± 10 °	
χ_{xz} 530 ± 60	$\mathcal{D}_1$ 3400 ± 300		
χ_{yz} 1300 ± 100	$\mathcal{D}_2$ 1475 ± 150		

Table 7.4: Values of the components of anisotropic part of magnetic suscep-tibility tensor and those of the Euler angles of Yb·**25** 1:1, calculated with the program PERSEUS, as explained in the text. Both reference systems are cen-tered on the paramagnetic ion (see figure 7.7).

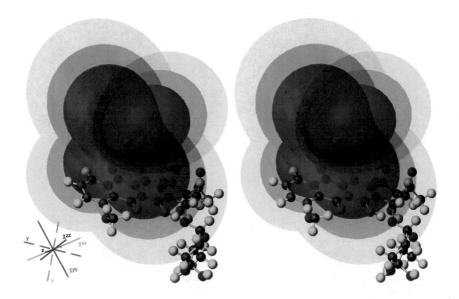

Figure 7.7: Stereoscopic view of experimentally determined iso-surfaces corresponding to pseudocontact shifts of ±20, ±10 and ±5 ppm, respectively, superimposed onto a ball-and-stick representation of the structure of Yb·**25**. Positive and negative pseudocontact shift values are indicated by blue and red colors, respectively. The Yb^{3+} ion is the yellow sphere located between the lobes. The principal axes system of the $\tilde{\chi}$ tensor and the molecular reference frame employed during the optimization are also displayed.

7.5 Detection of second complex Yb·26

We mentioned that the formation of a second species of Yb·**26** prevented the ^{1}H–NMR spectrum assignment by EXSY. In this sense, the new anthracycline **26** is probably similar to doxorubicin [17]. First, an evolution of UV-Vis CD spectrum of Yb·**26** 1:1 in methanol was observed (figure 7.8).

Secondly, both in the Lu^{3+} and Yb·**26** ^{1}H–NMR spectrum time-dependent changes were found. In particular, the variation of resonances of the free form of the compound **26**, found in the diamagnetic Yb·**26** spectrum, is indicative of conformational changes in the first sugar ring (a new C$_5$H peak appears in the time at 3.88 ppm, a new C$_2$H$_2$ multiplets appear at 1.81, 1.60 ppm and a new C$_6$H$_3$ signal appears at 1.19 ppm), whereas the second sugar ring undergoes minor modifications. On the other hand, great differences between the paragmagnetic zone of the spectra of the two Yb·**26** species were also evident (table 7.6).

The different exchange kinetics of this second species allowed, furthermore, the assignment of nine peaks of the bound form by EXSY (table 7.6). These new ^{1}H–NMR peaks, found for the second bound form of **26**, denote that the Yb^{3+}

^{1}H	δ^{pc}_{exp} (ppm)	δ^{pc}_{calc} (ppm)
C_1H	-13.1	-13.1
C_4H		-6.7
C_2H		-3.5
C_3H		-3.5
$C_{1'}H$		-6.7
$C1''H$		-2.6
C_7H	-11.9	-11.6
$C_{14}H_2$	-3.6	-3.4
$C_{5''}H$		-1.1
$C_{5'}H$		-4.4
$C_{3'}H$		-5.6
$C_{3''}H$		-1.3
$C_{4'}H$		-2.3
$C_{4''}H$		-0.8
$C_{10}H_2$	-41.2	-41.0
	-35.2	-35.0
C_8H_2	-11.5	-13.0
	-8.9	-8.7
$C_{2''}H_2$		-2.3
		-1.9
$C_{2'}H_2$	-5.7	-4.8
		-3.7
$C_{6'}H_3$		-2.3
		-1.1
		-1.9
$C_{6''}H_3$		-1.2
		-0.6
		-0.9

Table 7.5: Assigned experimental (δ^{pc}_{exp}) and calculated (δ^{pc}_{calc}) ^{1}H pseudocontact shifts of Yb·**26** 1:1.

ion is located on a different binding site. Moreover, beginning from 24 hours after solution preparation, Yb·**26** 1:1 in methanol exhibited a induced NIR CD band, centered at 975.5 nm (figure 7.8). This results show unambiguously that in the time a new complexation equilibrium establishes and it leads to conformational variation in **26**, as well as involvement of a second chiral coordination site for the ion, which probably should be the C_9OH, $C_{13}O$, $C_{14}OH$ system, absent in daunorubicin, but present in doxorubicin.

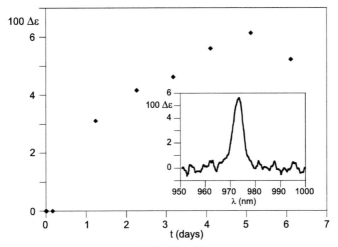

Figure 7.8: Time dependence of NIR CD of Yb·**26** 1:1 in methanol (4 mM) at 975.5 nm; in the square the NIR CD spectrum after 6 days is reported.

^{1}H δ (ppm)			
Yb·26 1:1 **(2nd bound form)**		**Yb·26 1:1** **(2nd free form)**	
-41.7	$(C_{14}H_2)$	8.3	(C_1H, C_4H)
-13.1	(C_8H)	7.9	(C_2H, C_3H)
-10.0	$(C_{10}H)$	5.01	$(C_{1'}H, C_7H)$
-9.6	(C_8H)	4.9	$(C_{1''}H)$
-7.2	$(C_{10}H)$	4.7	$(C_{14}H_2)$
-6.2	$(C_1H$ or $C_4H)$	4.4	$(C_{5''}H)$
-1.6		4.0	$(C_{3'}H)$
0.5	(C_7H)	3.9	$(C_{5'}H)$
2.4	$(C_1H$ or $C_4H)$	3.6	$(C_{3''}H, C_{4'}H, C_{4''}H)$
4.3	$(C_2H$ or $C_3H)$	3.3, 3.1	$(C_{10}H_2)$
4.6	$(C_2H$ or $C_3H)$	2.6, 2.4	(C_8H_2)
		2.0	$(C_{2''}H_2)$
		1.8, 1.6	$(C_{2'}H_2)$
		1.2	$(C_{6'}H_3)$
		1.2	$(C_{6''}H_3)$

Table 7.6: Chemical shifts (δ, ppm TMS refered) of second free and bound form of Yb·**26** 1:1 in CD_3O (known assignment are reported between round brackets).

7.6   Conclusions

The present study demonstrates the usefulness of the trivalet lanthanide ions, particularly of ytterbium, for the structural characterization of labile metal complexes with biomolecules through complementary use of optical techniques (UV-Vis absorption and CD, NIR CD), the ^{1}H-NMR spectroscopy and the numerical analysis of the NMR data. To this purpose, the isomorphism between the complexes of lanthanides and other of biologically more relevant cations can be effectively demonstrated by spectroscopic investigation. In this work, we carried out a detailed study on chemical properties both of the new anthracyclin **26** and its metal ion complexes. In a preliminary phase, we determined the stoichiometric ratio of the $Yb^{3+} \cdot$ **26** system by optical tritrations; particularly a predominat complex 1:1 both in methanolic and aqueous solution was found. Then, we verified the isomorphism of the Ca^{2+} and $Ln^{3+} \cdot$ **26** 1:1 complexes, detecting a correspondence between the variations of the CD, absorption and ^{1}H-NMR spectra of **26** following the coordination of Ca^{2+} or Ln^{3+}. During the specific study of the metal ions binding interaction of **26** we identified that a metal ion (La^{3+}, Yb^{3+}, Lu^{3+}, Ca^{2+}) induced deprotonation of the idroquinone group O-11, O-12 to give an anionic form, which yields a unique ion binding site in the Yb·**26** 1:1 complex, as shown by the analogy with the numerical analysis of the NMR data of daunorubicin. We showed also the usefulness of the NIR CD, that, together with the other spectrocopic techniques, contributed to the identification of the slow formation of a second stereodefined Yb·**26** complex, involving very probably the O-9, O-13, O-14 system as binding site. The similarities between the properties of **26** or its complex with Yb(III) in aqueous and in methanol solution suggest that the results presented in this work may be used for the understanding of the metal ion interactions with this new drug in physiological conditions. On the other hand, the structural similarity between the complexes of the two kinds of ions (Ca^{2+}, Ln^{3+}) confirms that the lanthanides and calcium adducts are isomorph. Additionally, in this work we verified that the lanthanide ions bind more tightly to the drug than Ca^{2+}.

Bibliography

[1] ARCAMONE, F.; CASSINELLI, G., *Curr. Med. Chem.* **1998**, 5, 391–419.

[2] ARCAMONE, F., *Doxorubicin Anticancer Antibiotics*; ACADEMIC PRESS: NEW YORK, 1981.

[3] WEISS, R. B.; SAROSY, G.; CLAGETT-CARR, K.; RUSSO, M.; LEYLAND-JONES, B., *Cancer Chemoter. Pharmacol.* **1986**, 18, 185–197.

[4] COURSEILLE, C.; BUSETTA, B.; GEOFFRE, S.; HOSPITAL, M., *Acta Crystallogr.* **1979**, B35, 764–767.

[5] LOWN, J. W., *Chem. Soc. Rev.* **1993**, 22, 165–176.

[6] PINDUR, U.; HABER, M.; SATTLER, K., *J. Chem. Educ.* **1993**, 70, 263–272.

[7] MINOTTI, G.; CAIRO, G.; MONTI, E., *FASEB J.* **1999**, *13*, 199–212.

[8] FEENSTRA, J.; GROBBEE, D. E.; REMME, W. J.; STRICKER, B. H. C., *J. Am. Coll. Card.* **1999**, *33*, 1152–1162.

[9] HÜSKEN, B. C. P.; DE JONG, J.; BEEKMAN, B.; ONDERWATER, R. C. A.; VAN DER VIJGH, W. J. F.; BAST, A., *Cancer Chemoter. Pharmacol.* **1995**, *37*, 55–62.

[10] CHEUNG, B. C. L.; SUN, T. H. T.; LEENHOUTS, J. M.; CULLIS, P. R., *Biochim. Biophys. Acta* **1998**, *1414*, 205–216.

[11] ARCAMONE, F.; ANIMATI, F.; CAPRANICO, G.; LOMBARDI, P.; PRATESI, G.; MANZINI, S.; SUPINO, R.; ZUNINO, F., *Pharmacology and Therapeutics* **1997**, *76*, 117–124.

[12] ANIMATI, F.; ARCAMONE, F.; BERETTONI, M.; CIPOLLONE, A.; FRANCIOTTI, M.; LOMBARDI, P., *J. Chem. Soc., Perkin Trans. 1* **1996**, 1327–1329.

[13] CANADA, R. G.; CARPENTIER, R. G., *Biochim. Biophys. Acta* **1991**, *1073*, 136–141.

[14] FIALLO, M. M. L.; GARNIER-SUILOEROT, A., *Biochemistry* **1986**, *25*, 624–930.

[15] MCLENNAN, I. J.; LENKINSKI, R. E., *J. Am. Chem. Soc.* **1984**, *106*, 6905–6909.

[16] BERALDO, H.; GARNIER-SUILLEROT, A.; TOSI, L., *Inorg. Chem.* **1983**, *22*, 4117–4124.

[17] WEI, X.; MING, L. J., *Inorg. Chem.* **1998**, *37*, 2252–2262.

[18] MING, L. J.; WEI, X., *Inorg. Chem.* **1994**, *33*, 4617–4618.

[19] HAJ-TAYEB, H. B.; FIALLO, M. M. L.; GARNIER-SUILLEROT, A.; KISS, T.; KOZLOWSKI, H., *J. Chem. Soc., Dalton Trans.* **1994**, 3963–3968.

[20] LENKINSKI, R. E.; SIERKE, S.; VIST, R., *J. Less-Common Metals* **1983**, *94*, 359–365.

[21] PAPAKYRIAKOU, A.; ANAGNOSTOPOULOU, A., "PREPARATION OF URANYL COMPLEXES WITH ANTHRACYCLINES AND THEIR INTERACTION WITH DNA", IN *Book of Abstracts, 34th Conference on Coordination Chemistry* EDINBURGH, 2000 .

[22] EVANS, C. H., *Biochemistry of the Lanthanides;* PLENUM PRESS: NEW YORK, 1990.

[23] COTTON, S., *Lanthanides and Actinides;* OXFORD UNIVERSITY PRESS: NEW YORK, 1991.

[24] KEMPLE, M. D.; RAY, B. D.; LIPKOWITZ, K. B.; PRENDER-GAAST, F. G.; RAO, B. D. N., *J. Am. Chem. Soc.* **1988**, *110*, 8275–8287.

[25] REUBEN, J., *Naturwissenschaften* **1975**, *62*, 172–178.

[26] BERTINI, I.; LUCHINAT, C., *Solution NMR of paramagnetic molecules. Applications to metallobiomolecules and models;* ELSEVIER: AMSTERDAM, 2001.

[27] RICHARDSON, F., *Inorg. Chem.* **1980**, *19*, 2806–2812.

[28] MCLENNAN, I. J.; LENKINSKI, R. E., *J. Am. Chem. Soc.* **1984**, *106*, 6905–6909.

[29] MING, L.-J.; WEI, X., *Inorg. Chem.* **1994**, *33*, 4617–4618.

[30] WEI, X.; MING, L.-J., *Inorg. Chem.* **1998**, *37*, 2255–2262.

[31] ARCAMONE, F. *et al.* , *In Vivo J. National Cancer Inst.* **1997**, *89*, 1217–1223.

[32] PRATESI, G.; DECESARE, M.; CASERINI, C.; PEREGO, P.; DALBO, L.; SUPINO, R.; BIGIONI, M.; MANZINI, S.; IAFRATE, E.; SALVATORE, C.; CASAZZA, A. M.; ARCAMONE, F.; ZUNINO, F., *Clinical Cancer Res.* **1998**, *4*, 2833–2839.

[33] CIRILLO, R.; SACCO, G.; VENTURELLA, S.; BRIGHTWELL, J.; GIACHETTI, A.; MANZINI, S., *J. Cardiovascular Pharmac.* **2000**, *35*, 100–108.

[34] MENOZZI, M.; VALENTINI, L.; VANNINI, E.; ARCAMONE, F., *J. Pharm. Sci.* **1984**, *73*, 766–770.

[35] CHAIRES, J. B.; DATTAGUPTA, N.; CROTHERS, D. M., *Biochemistry* **1982**, *21*, 3957–3962.

[36] MARTIN, S. R., *Biopolymers* **1980**, *19*, 713–721.

[37] MCLENNAN, I. J.; LENKINSKI, R. E.; YANUCA, Y. A., *Can. J. Chem.* **1985**, *63*, 1233–1238.

Chapter 8

Interaction of YbDOTA with γ-cyclodextrin

A very relevant application of NIR-CD spectroscopy can be found in the study of effects induced following interactions with chiral biological molecules. [1–4] To make an example, it is well-recognized that several contrast agents for MRI do interact with plasma proteins, [5–8] which in turn are often highly stereodiscriminating [9] and can induce preferential enantiomeric conformations. [10, 11] Surprisingly, no systematic study on the role of the stereochemistry of contrast agents (CA) could be found in the literature. One possible reason is the lack of analytical tools for sensing the configuration around the metal ion and CD of lanthanide ions and in particular of Ytterbium can provide the missing information.

8.1 Protein-bound contrast agents

Two main reasons have stimulated research in the field of the interaction between biological macromolecules and contrast agents for MRI:

1. the effectiveness of a gadolinium-based constrast agent, that is its ability to enhance the longitudinal water relaxation, is greatly determined by its rotational correlation time τ_r, an increase in this parameter resulting in shorter T_{1w} values [5]. Protein binding can therefore couple the strong chelation of the metal ion provided by the multidentate ligand with a slow molecular tumbling of the protein, featuring high relaxivity with reduced administered doses of CA [6, 12];

2. the interaction of the chelate with a particular biological substrate can promote an higher residence time of the complex in a determinate tissue, providing a better compartimentalisation of the CA and an increased resolution of the imaging [5].

Such peptide-CA conjugation has been achieved:

- either through a covalent linking of the gadolinium chelates to amino acid residues of the protein [12, 13]

- or by promoting a non-covalent binding between slowly tumbling macromolecules and suitably functionalized complexes (e.g. containing hydrophobic side chains, like the β-benzyloxy-α-propionic substituent in the commercial GdBOPTA [8, 14]).

8.2   Cyclodextrins and lanthanide complexes

Cyclodextrins (α, β and γ CDs) are cone-like shaped molecules wiyh an hydrophobic cavity made up by α-1,4-linked D-glucopyranoside units (6, 7 or 8, respectively) [15, 16].

They have been the object of several NMR investigations concerning the thermodynamic stability and the structure of host-guest compounds with ions, organic products, biological drugs (see for example [17, and references therein]). In these analysis, a double kind of attention has been drawn to CDs and their inclusion compounds:

- as "drug delivery systems" and

- as enzyme model systems for probing molecular recognition processes in biological systems.

In the field of magnetic imaging, the first point has been diffusely developed in the case of Gd(III) chelates endowed with aromatic-containing side arms [18, 19], where the occurrence of an interaction could be exploited to alter the *in vivo* biodistribution of the contrast agent, increasing at the same time its relaxivity.

Nevertheless, the possibility of using CDs as "carriers" for complexes like LnDOTA$^-$ or LnDOTP^{5-} has been recently shown by Geraldes *et al.* to be an unpracticable route [2, 3]. Though an adduct was revealed between γ-CD and TmDOTA$^-$ and TmDOTP^{5-} and the analysis of the ^{1}H-NMR shifts induced by the lanthanide on the sugar protons allowed a description of the resulting structure, the low stability constants of the inclusion compounds led to the conclusion that interaction of contrast agents like GdDOTA or GdDOTP with γ-CD is not an useful approach to increase their efficacy.

On the contrary, the second point can deserve additional considerations, since cyclodextrins offer the possibility to model a binding which is likely to be *specific* and *stereocontrolled*.

In the previous chapters (3 and 5), we described the effect brought about by chiral groups **on** the ligand on the structural conformations and solution dynamics of the various Yb^{3+} complexes. When taking into account an *achiral* complex like YbDOTA, we saw anyway that all the different structures involved in its conformational equilibria are endowed with a defined chirality; it some kind of intermolecular binding happens to be preferential for one of the different conformations adopted, complexation with high molecular aggregates or macromolecules could be exploited to force the global equilibrium network toward a particular conformation. This "molecular host" will therefore be playing (in a *non-covalent*) fashion, the same role of a *covalent* functionalization of the side arm.

The fine mechanisms by which such chiral recognition work will include coperative effects of electrostatic or hydrophobic interactions, hydrogen bonds and structural influences; their ultimate understanding is therefore essential in the optimization of the efficiency of lanthanide complexes as protein-bonding contrast agents in structural studies.

As a first example, in the case of YbDOTA, it should be checked whether the binding with γ-CD occurs:

- preferentially for one of the two *p*- or *n*-forms, thus altering the equilibrium populations of the two coordination polyhedra;

- preferentially for one of the two Λ or Δ distortions of the coordination cage (the γ-CD would act as a chiral resolving agent [20] for the enantiomeric equilibrium mixture of the metal complex, as, for example, D-(RR)tartrate does towards octahedral transition metal complexes [21] or N-methyl-D-glucamine towards LnDOTA itself [1]).

Both points have been neglected in the analyses of Geraldes *et al.*, who explicitly assumed that the binding to the CD does not affect the isomer equilibrium of TmDOTA [3]. The NMR detection of such effects is actually quite difficult, since the low stability constant of the inclusion compounds is translated in negligible variations of the integrals of the relevant NMR signals of the DOTA nuclei and in isochronous environments for the distereomers eventually formed. On the contrary, this situation can be solved exploiting the properties of the f electrons of the Ln(III) ions: as we saw, this behaviour is in turn strongly echod in the electronic structure of the ion, which becomes itself chiral. Comparing the spectrum with that of model chiral compounds (as those studied in chapter 3), the signs and magnitudes of the the induced CD bands (ICD) can be related to a specific distortion (Λ or Δ) of the coordination polyhedron and also binding constants could in principle be calculated. Moreover, provided that n- and p-forms of the Ln^{3+} complex feature slightly different (discernible) patterns in the electronic CD, not only the helicity but also the whole shape of the coordination polyhedron could be described by such an analysis.

8.3   NIR-CD of YbDOTA with γ-CD

In the present section, the interaction between YbDOTA and γ-CD was studied by screening the NIR-CD of Yb^{3+}. The low stability constant which is expected for the adduct can be balanced by the high g-factor associated with the $^2F_{7/2} \rightarrow ^2F_{5/2}$ transition of this ion (see section 1.2.4), which is therefore likely to be the most suitable probe for such kind of analysis.

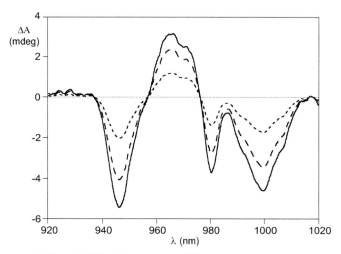

Figure 8.1: Induced NIR-CD spectra of YbDOTA with γ-CD in water solution ($[\gamma CD]$=0.050 mM; $\rho = \frac{[YbDOTA]}{[\gamma CD]}$=1 (dotted line), 2 (dashed line) and 4 (continuous line)).

To a water solution of YbDOTA, increasing amounts of γ-CD were progressively added. The spectra of figure 8.1 were recorded at molar ratios $\rho = \frac{[YbDOTA]}{[\gamma CD]}$ of 1, 2 and 4, respectively.

Assuming comparable binding constants for YbDOTA and TmDOTA (β_1=0.33 M^{-1}, β_2=3.4 M^{-1}, according to ref. [3]), the relative proportions of the species containing bound YbDOTA can be extimated (table 8.1): only one major species is therefore likely to contribute to the ICD spectrum.

$[\gamma CD]_{tot}$ = 50 mM	$\rho = 1$	$\rho = 2$	$\rho = 4$
[YbDOTA]	49.1 mM	98.2 mM	196.5 mM
[γCD]	48.9 mM	47.9 mM	46.1 mM
[YbDOTA $\cdot \gamma CD$]	0.8 mM	1.6 mM	3.0 mM
[(YbDOTA)$_2 \cdot \gamma CD$]	0.1 mM	0.2 mM	0.4 mM

Table 8.1: Concentration of the different species in solution in the NIR-CD experiments performed with different molar ratios ρ between γ-CD (50 mM) and YbDOTA (50, 100 and 200 mM), assuming stability constants β_1=0.33 M^{-1} and β_2=3.4 M^{-1} as for TmDOTA [3].

8.4   Conclusions

While the NIR-CD spectrum of YbDOTMA (pag. 89) acts as a standard for a *p*-type molecule, the lacking of a reference spectrum for a *n*-type carboxylate complex, anyway, prevented us so far from a straightforward analysis of the recorded pattern.

Anyway, it is noteworthy that:

- an interaction between YbDOTA and γ-CD was sensed;

- a stereoinduction follows the interaction: this means that chirality is transferred from a site in the *host* to the overall twist of the coordination polyhedron of the Yb^{3+} *guest*.

A more complete study is currently in progress.

Bibliography

[1] AIME, S.; BOTTA, M.; CRICH, S. G.; TERRENO, E.; ANELLI, P. L.; UGGERI, F., *Chem. Eur. J.* **1999**, *5*, 1261–1266.

[2] SHERRY, A. D.; ZARZYCKI, R.; GERALDES, C. F. G. C., *Magn. Reson. Chem.* **1994**, *32*, 361–365.

[3] ZITHA-BOVENS, E.; VAN BEKKUM, H.; PETERS, J. A.; GERALDES, C. F. G. C., *Eur. J. Inorg. Chem.* **1999**, 287–293.

[4] AIME, S.; BETTINELLI, M.; FERRARI, M.; RAZZANO, E.; TERRENO, E., *Biochim. Biophys. Acta* **1998**, *1385*, 7–16.

[5] AIME, S.; BOTTA, M.; FASANO, M.; TERRENO, E., *Chem. Soc. Rev.* **1998**, *27*, 19–29.

[6] AIME, S.; BOTTA, M.; FASANO, M.; CRICH, S. G.; TERRENO, E., *J. Biol. Inorg. Chem.* **1996**, *1*, 312–319.

[7] KIRCHIN, M. A.; PIROVANO, G. P.; SPINAZZI, A., *Invest. Radiology* **1998**, *33*, 798–809.

[8] CAVAGNA, F. M.; MAGGIONI, F.; CASTELLI, P. M.; DAPRÁ, M.; IMPERATORI, L. G.; LORUSSO, V.; JENKINS, B. G., *Invest. Radiology* **1997**, *32*, 780–796.

[9] ELST, L. V.; CHAPELLE, F.; LAURENT, S.; MULLER, R. N., *J. Biol. Inorg. Chem.* **2001**, *6*, 196–200.

[10] KRAGH-HANSEN, U., *Pharmacol. Rev.* **1981**, *33*, 17–53.

[11] SALVADORI, P.; BERTUCCI, C.; ASCOLI, G.; UCCELLO-BARRETTA, G.; ROSSI, E., *Chirality* **1997**, *9*, 495–505.

[12] CARAVAN, P.; ELLISON, J. J.; McMURRY, T. J.; LAUFFER, R. B., *Chem. Rev.* **1999**, *99*, 2293–2352.

[13] LIU, S.; EDWARDS, D. S., *Bioconj. Chem.* **2001**, *12*, 7–34.

[14] UGGERI, F.; AIME, S.; ANELLI, P. L.; BOTTA, M.; BROC-CHETTA, M.; DE HAËN, C.; ERMONDI, G.; GRANDI, M.; POLI, P., *Inorg. Chem.* **1995**, *34*, 633–642.

[15] SAENGER, W., *Angew. Chem. Int. Ed. Engl.* **1980**, *19*, 344–362.

[16] SZEIJTLI, J., *Cyclodextrins and their inclusion complexes;* AKADEMIAI KIADO: BUDAPEST, 1982.

[17] UDACHIN, K. A.; RIPMEESTER, J. A., *J. Am. Chem. Soc.* **1998**, *120*, 1080–1081.

[18] AIME, S.; BOTTA, M.; PANERO, M.; GRANDI, M.; UGGERI, F., *Magn. Reson. Chem.* **1991**, *29*, 923–927.

[19] AIME, S.; BOTTA, M.; FRULLANO, L.; CRICH, S. G.; GIOVEN-ZANA, G. B.; PAGLIARIN, R.; PALMISANO, G.; SISTI, M., *Chem. Eur. J.* **1999**, *5*, 1253–1260.

[20] PARKER, D., *Chem. Rev.* **1991**, *91*, 1441–1457.

[21] MIZUTA, T.; TOSHITANI, K.; MIYOSHI, K.; YONEDA, H., *Inorg. Chem.* **1990**, *29*, 3020–3026.

Conclusions

In this thesis we describe an original attempt to combine different approaches employing NMR and electronic spectroscopy for the structural and dynamic characterization of chiral rare earth chelates in solution. The molecules studied in the various chapters constitute a wide set of stable complexes important in MRI as contrast agents, in bioluminescence studies as highly emitting probes and in organic and bioorganic synthesis as chiral auxiliaries or intermediates, artificial nucleases and stereoselective catalysts.

^{1}H and ^{13}C NMR analysis exploits the constraints provided by the paramagnetic nature of the lanthanide ions and allows the description of the structure and dynamics of the coordination cage, that is of the geometries and the possible motions involving the ligands wrapped around the metal ion.

Electronic ff transition originating from the lanthanide ions are electrically forbidden, thus endowed with very small absorption coefficients. The presence of chiral centers on the ligand, though, produces a distortion of the environment surrounding the metal, reflected in the appearance of bands with high dissymmetry factors in the CD spectrum. The resulting signals can be recorded on suitable dichrographs with relative ease, in contrast to a somewhat common belief, and are sensitive to the crystal field in a way more explicit than NMR. We show how transition levels and ff band intensities are determined by the coordination number, the position of the coordinating groups and the resulting tilt angle of the coordination polyhedron. Such a stereochemical description, crucial for understanding the solution chemistry of these systems, cannot be provided by the NMR study, which reports only indirectly the effects of the donating atoms on the central cation.

Ytterbium is the most widely used lanthanide ion in our work: its absorption in a spectral region (NIR) rather devoid of contribution from organic chromophores and are endowed with very favorable dissymmetry factors. Furthermore, owing to its particular electronic configuration, the geometric information coming from NMR is easier to evaluate and at the same time the different contribution of the two different spectroscopies can be combined in a simpler way.

In the different cases, we demonstrate how NIR-CD can join and complete NMR data in several different fashions:

- for completely qualitative analysis: the electronic ff transitions enable us to get information on the equilibrium dynamics of a given complex, that

is to investigate the number of species in solution and the thermodynamics of their interconversion, in cases where other spectroscopic means do
not allow any conclusion (e.g.: severe broadening of the lines in the NMR
spectrum; lack of other chromophoric units in the systems; ...);

- for structural determinations through empirical correlations: the analysis of the CD pattern is used to probe the helicity of the coordination
  polyhedron provided by the chiral ligand wrapped around the rare earth
  ion. The chiroptical spectroscopy provides the information missing from
  NMR analysis and allows one to draw empirical correlations between
  structurally similar chelates;

- semi-empirical study: for a fully characterized system (Yb·**2**), the NIR-
  CD spectrum shows several well-resolved transitions around 980 nm with
  very high dissymmetry factors, which can be tentatively assigned. A
  fully theoretical independent systems model for the ff transition, is able
  to reproduce the experimental CD spectrum, having as the input the
  solution structure determined from NMR data. This is the first attempt
  to a non-empirical approach to the problem of the NIR transitions of this
  ion.

Magnetic and optical parameters used in describing the behavior of lanthanide
ions show at the same time a strong correspondence and a useful complemen-
tarity in elucidating the nature of their systems in solution, in the case of stable
and inert complexes and for labile adducts as well. Our strategy allowed us,
in the different investigations, to achieve:

- information on the **metal cation** itself and on its degree of accessibility,
  which is a crucial step in all the functions performed by these molecules
  in solution (from organic catalysis to contrast enhancements in MRI);

- information on the **ligand**, when the lanthanide ion is used as a probe
  to define the stereochemistry of coordinated molecules (e. g. in the
  determination of the chirality of 1,2-diols or in the mapping of the metal
  binding site of a drug);

- information on the overall lanthanide coordination and on the global
  conformation of its geometry, which can modulate the association of the
  complex as a whole towards a substrate (e. g. in organic or bioorganic
  reaction).

The results show how the concerted application of NMR and NIR-CD spectroscopy can allow a deeper understanding of the coordination properties of
these attractive complexes and, more in general, of several facets of the solution
chemistry of lanthanides.

Part IV

Appendix

Appendix A

Experimental

A.1 Molecules

The anionic YbDOTA complex (Yb-**1**) was prepared as a sodium salt according to Desreux [1]. Its water and methanol solutions were prepared at concentrations between 10 and 160 mM.

The anionic homochiral and enantiopure YbDOTMA complex (Yb-**2**) as a methylglucammonium salt was a gift from Bracco Italy S.p.A.

Cationic enantiopure Yb·(R)-**3** and Yb·(S)-**4** complexes were obtained as their trifluoromethanesulphonate salts as previously described [2–4].

Samples of cationic enantiopure Yb·(R)-**3**, Yb·(S)-**3**, Pr·(R)-**4**, Eu·(R)-**4**, Dy·(R)-**4** and Yb·(R)-**4** compexes were provided by prof. D. Parker and R. S. Dickins (University of Durham, U.K.).

Cationic enantiopure Yb·(S)-**5**, Yb·(S)-**6** and Yb·(S)-**7** complexes were synthetized by prof. B. Feringa (University of Groningen, the Netherlands).

Enantiopure (S)-THP (S)-**8** was synthetized from cyclen and S-propilene oxide as previously reported [5] and the cationic Yb·(S)-THP, Ce·(S)-THP and Lu·(S)-THP complexes were prepared as their trifluoromethanesulphonate salts following the literature procedure.

Enantiopure YbNa$_3$BINOL$_3$ (**9**) and LuNa$_3$BINOL$_3$ (**10**) were synthetized following, with same modifications, the recipe for the analogue Pr^{3+}, Nd^{3+} and Eu^{3+} complexes [6, 7], reacting anhydrous Yb(OTf)$_3$ and LuCl$_3$ with vigorous stirring for 20 hours at $50°C$ with (S)-1,1'-bi-(2-naphthol) (3 eq.), NaO^{t}Bu (6 eq.) and water (6 eq.) in THF. Crystals of the complexes of the right dimension for X-ray analysis were obtained upon slow cooling to room temperature of the solution.

Enantiopure YbK$_3$BINOL$_3$ (**11**) was obtained modifying the procedure of ref. [8], reacting Yb(OTf)$_3$ with vigorous stirring for 20 hours at $50°C$ with (S)-1,1'-bi-(2-naphthol) (3 eq.), KHMDS (6 eq.) and water (1 eq.) in THF.

Yb(fod)$_3$ **12** was purchased from Aldrich.

Diols **13**, **15** and **17** were commercially available (Aldrich and Fluka). Diols **14**, **16** and **19–21** have been synthetized according to the Sharpless procedure

[9–12] in the laboratories of prof. Pini, University of Pisa, who kindly provided samples of the enantiopure products.

The disaccharide anthracycline MEN 10755 (**25**) was supplied from Menarini-Firenze, Italy.

Metal cation clorides or triflates are anhydrous commercial products (purity degree $\geq$ 97%).

The tris(hydroxymethyl)aminomethano buffer (TRIS) is a Carlo Erba product. The used solvents are HPLC grade.

Solutions of complexes Ln·**1**–Ln·**11** for NMR and NIR-CD spectroscopy were prepared by dissolving the solid complexes in methanol-d_4, dimethylsulpho-xide-d_6, acetonitrile-d_3 and tetrahydrofuran-d_8 or the corresponding non-deuteriated solvents in the range 10-100 mM.

To a sample of YbDOTMA (Yb-**1**) in D_2O, in order to carry out NMR and NIR-CD measurement simultaneously, a 100-fold excess of KF (Fluka) was added in order to further reduce the concentration of the minor form.

To the water solutions of Yb-(S)-THP (Yb·(S)-**8**), 0.1 M NaCl was added to mantain constant ionic strength; the pD (or pH), initially 7, was varied in the range 2.3-9.1 by the addition of 0.1 M HCl and 0.1 M NaOH.

Solutions for ^{1}H-NMR and NIR-CD spectroscopy with **25** were about 5-10 mM in pH 7.6 aqueous TRIS buffer or unbuffered methanol.

Solutions pHs (uncorrected values) were measured at 25°C by use of an Orma digital pHmeter equipped with temperature compensation probe.

A.2   NMR data

The NMR spectra were recorded on a Varian VXR 300 Spectrometer operating at 7.1 T, equipped with a VT unit stable within 0.1° C.

Chemical shifts were referenced with respect to tert-butyl alcohol as internal standard; in the spectra of Yb(fod)$_3$, the chemical shifts are referred to the residual signal of CHCl$_3$ at 7.24 ppm.

The 90° ^{1}H pulselength ranged between 13 and 14.5 μs. The recycle delay in all experiments was carefully chosen to largely exceed $5 \cdot T_1$ of all proton resonances of the complexes and ensuring instrumental recovery ($\simeq$ 5 s for diamagnetic samplex, $\simeq$ 0.5 s for strongly paramagnetic complexes).

Steady-state NOE spectra were recorded with 4K up to 32K transients after 0.5 s cw-presaturation of the relevant signals.

T_1 measurements were made by using the standard inversion-recovery sequence and three parameter single exponential fits of the signal recoveries. T_2 values were calculated from the line widths. In the case of YbNa$_3$BINOL$_3$, the ^{1}H-NMR spectrum was deconvoluted by the built-in routine in order to take into account incompletely resolved multiplicities.

TPPI NOESY [13, 14], TOCSY [15] and DQF-COSY (COSY in the text) [16] spectra were acquired by conventional sequences. Mixing times were varied between 30 and 100 ms.

Since zero quantum coherences have an antiphase character, they cannot contribute neither in integrated nor in point amplitude in systems with a linewidth largely exceeding the scalar couplings, thus no attempt was made to suppress them. DQF-COSY (COSY in the text) spectra were acquired by a conventional sequence.

In a typical experiment, 2048 points were collected for each of the 512 t1 increments, covering a spectral window of 50000 Hz in two dimensions. The matrices were zero-filled to $2K \times 2K$ and weighted by Lorentz-to-Gauss or shifted sinebell functions in both dimensions.

HETCOR spectra of YbNa$_3$BINOL$_3$ and LuNa$_3$BINOL$_3$ were recorded using a standard pulse sequence [17] and were realized over a window of 20,000 Hz in the direct dimension and 18,248 Hz in the indirect one. A total of 256 fids were collected, of 128 repetitions each. The correlation constant was fixed to 140 Hz. A 45-60° shifted squared-sine-bell apodization function was applied to both dimensions prior to Fourier transformation.

In the case of YbDOTMA, since all quantitative analyses were performed only on the methyl resonances (within 12000 Hz from the transmitter), no correction for finite pulselength was used.

No baseline correction was applied and the 1- and 2-D integrals were evaluated with the VNMR software with reference to the area immediately adjacent to each peak.

Uncertainties in the chemical shift and relaxation measurements were $\pm$ 1-2% and $\pm$5-10%, respectively.

A.3 UV-absorption and CD spectra

The UV-Vis absorption and CD spectra were detected on a Varian Cary 4/E spectrophotometer and a Jasco J-600 dichrograph, respectively. Path lengths were varied between 0.01 and 0.1 cm; the band-passing was 1.0 nm; typical time costants were about 4 sec, speed 20 nm/min. All the spectra of the chiral diol adducts were obtained directly after mixing the compounds in CCl$_4$ anhydrified on molecular sieves 4 Å and did not evolve with time.

A.4 NIR-absorption and CD spectra

The NIR absorbance spectra were recorded with a Perkin Elmer Lambda 9 spectrophotometer operating between 250 and 1200 nm. The NIR-CD spectra were recorded with a spectropolarimeter JASCO J-200D coupled to a JASCO DP-500N data processor; operating between 750 and 1350 nm, modified with a tandem Si/InGaAs detector with a dual photomultiplier amplifier [18]. All the spectra are obtained after 1 to 4 acquisition at the rate of 5-50 nm/min, time constant 0.5-8 sec, band-passing 2-3.4 nm, sensitivity 1-100 mdeg/cm.

The path length was 1 cm for all samples, with the exception of Yb(fod)$_3$ (10 cm).

Only for the more diluted solutions of Yb·1–Yb·11 or for the more labile diol adduct of section 6.1 and 6.3, up to 64 acquisitions were averaged to improve the signal-to-noise ratio.

The spectra were recorded via digital read-out equipment connected directly to the instrument. As the basis for resolution and curve fitting, performed with a home-built routine on KaleidaGraph (Abelbeck Software) software package, a Lorentzian function of the form:

$$y = \sum_{i=1}^{m} \frac{\Delta \epsilon_i^{max}}{\left(\frac{x - \tilde{\nu}_i}{\Delta \tilde{\nu}_i}\right)^2 + 1} \tag{A.1}$$

was used, where m is the number of components of the band, $\Delta \epsilon_i^{max}$ is the peak height, $\tilde{\nu}_i$ the energy at which the latter is observed and $2\Delta \tilde{\nu}_i$ is the peak half-height width.

All the spectra of the chiral diol adducts were obtained directly after mixing the compounds in CCl$_4$ anhydrified on molecular sieves 4 Å and did not evolve with time.

A.5   Luminescence measurements

Luminescence emission spectra and radiative rate constants characterising the decay of emission from the Yb excited state were measured by prof. D. Parker using the apparatus described in ref. [19].

A.6   Molecular optimization

The conformation of the sugar portion of **25** was optimized on the basis of NOE costraints by means of a MM2 simulation through a commercially available molecular mechanics software (CS Chem3D Pro, version 3.5.2, CambridgeSoft Corporation).

The values and orientation of ytterbium magnetic anisotropy tensor in the various complexes, the layout of the side arms of complexes Yb·1–Yb·8, the position of the Yb^{3+} ion in the 1:1 complex with butanediol **13** and **25** and in the heterobimetallic complex with BINOL, were determined on the basis of pseudocontact chemical shift and relaxation data by means of a newly designed FORTRAN routine, "PERSEUS", which will be discussed in the following paragraph.

A.6.1   The routine "PERSEUS"

As shown in secton 2.1.2 and 2.2.3, given a molecular geometry, the pseudocontact shifts of the nuclei and their relaxation times are completely determined by the magnetic susceptibility anisotropy tensor ($\tilde{\chi}$) values and by the electronic and rotational correlation times τ_s and τ_r. Therefore, it is possible to use the data obtained from NMR as restraints in the structure determination of the paramagnetic complex in solution.

The FORTRAN program PERSEUS (Paramagnetic Enhanced Relaxation and Shift for Eliciting (Elucidating) Ultimate Structures, see figure A.1) locates one or more (non-coupled) paramagnetic ions within a given molecule by means of the combined constraints given by relaxation times and pseudocontact induced shifts.

It constitutes an extension and a refinement of the simple algorithms which have been used in the interpretation of paramagnetic NMR data on small complexes (see for example [20] or a review in [21]).

It allows the simultaneous positioning of the metal ions within the ligand, the calculation of the $\tilde{\chi}$ values and an optimization of the molecular geometry, since one or more portions of the ligand are allowed to rotate, as well, around suitably defined axes.

If the pseudocontact shifts of at least 8 nuclei are known in a rigid portion of the molecule, it is possible to determine the 5 independent components of the $\tilde{\chi}$ tensor and the coordinates of each paramagnetic ion, as well, in the portion of system at issue. The principal values and the orientation of the $\tilde{\chi}$ tensor are then determined by diagonalization. This procedure has been carefully described in section 2.1.2 and makes no hypotesis on the direction of the principal axes of $\tilde{\chi}$ tensor or on the spatial symmetry of the complex.

Starting from a given geometry for the ligand, a set of pseudocontact shifts δ_i^{pc} is calculated according to:

$$\delta_i^{pc} = \sum_n \frac{1}{12\pi r_{in}^5} \left\{ (\chi_{zz\,n} - \frac{\chi_{xx\,n} + \chi_{yy\,n}}{2}) \left[3(z_i - z_n)^2 - r_{in}^2 \right] + \right.$$

$$\frac{3(\chi_{xx\,n} - \chi_{yy\,n})}{2} \left[(x_i - x_n)^2 - (y_i - y_n)^2 \right] + 6\chi_{xy\,n}(x_i - x_n)(y_i - y_n) +$$

$$\left. 6\chi_{xz\,n}(x_i - x_n)(z_i - z_n) + 6\chi_{yz\,n}(y_i - y_n)(z_i - z_n) \right\} \quad \text{(A.2)}$$

where x_i, y_i and z_i are the nucleus coordinates, x_n, y_n and z_n are the coordinates of the n-th paramagnetic ion, $\chi_{lm\,n}$ its magnetic suscptibility (lm) components and:

$$r_{in} = \sqrt{(x_i - x_n)^2 + (y_i - y_n)^2 + (z_i - z_n)^2}.$$

For each paramagnetic ion, the three coordinates and the five components of the magnetic anisotropic susceptibility tensor of eq. A.2 are continuously varied by means of the Gauss-Marquardt algorithm in order to minimize the value of the agreement factor R between calculated δ_i^{calc} and experimental δ_i^{exp} values over the whole set of spin:

$$R = \sqrt{\frac{\sum_i (\delta_i^{calc} - \delta_i^{exp})^2}{\sum_i (\delta_i^{exp})^2}} \tag{A.3}$$

The coordinates x_i, y_i and z_i can be also progressively changed during the simulation, while a determined rotation is accomplished in a portion of the ligand.

The $\mathcal{D}_1$ and $\mathcal{D}_2$ parameters of equation 2.22 of pag. 29, which represent the components of the $\tilde{\chi}$ tensor in its principal system, can then be obtained by a simple diagonalization; at the same time, the three Euler angles, which determine the orientation of the principal axes relative to the reference molecular system, are calculated as well.

If relaxation data are available, for each geometry the quantity:

$$\frac{1}{T_{1i}^{calc}} = \sum_n \frac{C_n}{r_{in}^6} \tag{A.4}$$

is calculated, as well, and the generalized constants C_n (of the order of 10^5 s·Å^{-6} for lanthanides) are optimized, in order to minimize the *extended* agreement factor R':

$$R' = \sqrt{\frac{\sum_i (\delta_i^{calc} - \delta_i^{exp})^2}{\sum_i (\delta_i^{exp})^2} + \frac{\sum_i (\frac{1}{T_{1i}^{calc}} - \frac{1}{T_{1i}^{exp}})^2}{\sum_i (\frac{1}{T_{1i}^{exp}})^2}} \tag{A.5}$$

As described in section 2.2 and in particular at pag. 40, for molecules in the fast-motion regime, the value of the constant C is:

- a linear combination of τ_s and τ_r, if T_1^{-1} values are employed in which both the pseudocontact and Curie contributions are present,

- proportional to τ_r only, if the Curie contribution alone can be estimated from the difference $T_2^{-1} - T_1^{-1}$ (equation 2.58).

Exploiting the powers of Medusa's head, which transformed into stone whoever looked in her eyes, the greek hero Perseus (fig. A.1) was eventually able to save his mother Danaes and to free Andromeda, his future wife [22, part II, chapter 4]. It appears straightforward that similar deeds are accomplished by our constraining routine, which reduces the flexibility of the target molecule by means of the experimental NMR data.

A.7   Other experimental techniques

Solutions for ESI-MS spectrometry were prepared by dissolving the solid complex in MeOH, water and acetonitrile or the corresponding deuteriated solvents in the range 10-100 nM and were delivered to the mass spectrometer by

Figure A.1: Perseus lifting Medusa's head (cartoon from Canova's sculpture, Roma, Musei Vaticani).

continuous infusion by a syringe pump at 5 ml/min. Ionspray mass spectra were performed on a Perkin-Elmer Sciex API III plus triple quadrupole mass spectrometer (Sciex C., Thornhill, Ont., Canada) equipped with an API ion source and an articulated ionspray interface. The ESI spectra were obtained under the following experimental conditions: ionspray voltage, 5.5 kV; orifice voltage, 35 V; scan range, m/z 200-1500; resolution above 1 Da; scan rate as appropriate. The nature of the species observed was assigned on the basis of the isotopic pattern and from the fragmentation profile in MS-MS experiments. ESI product ions were produced by collision-induced decomposition (CID) of selected precursor ions in the collision cell of the PE Sciex API III plus and mass-analyzed using the second analyzer of the instrument. Other experimental conditions for the CID included: collision energy, 20 eV; collision gas thickness, $250 \cdot 10^{13}$ molecules/cm^2; scan range was variable, depending on the m/z value of the selected precursor ion.

Bibliography

[1] DESREUX, J. F., *Inorg. Chem.* **1980**, *19*, 1319–1324.

[2] DICKINS, R. S.; HOWARD, J. A. K.; MAUPIN, C. L.; MOLONEY, J. M.; PARKER, D.; RIEHL, J. P.; SILIGARDI, G.; WILLIAMS, J. A. G., *Chem. Eur. J.* **1999**, *5*, 1095–1105.

[3] DICKINS, R. S.; HOWARD, J. A. K.; LEHMANN, C. W.; MOLONEY, J.; PARKER, D.; PEACOCK, R. D., *Angew. Chem. Int. Ed. Engl.* **1997**, *36*, 521–523.

[4] DICKINS, R. S.; HOWARD, J. A. K.; MOLONEY, J. M.; PARKER, D.; PEACOCK, R. D.; SILIGARDI, G., *Chem. Comm.* **1997**, 1747–1748.

[5] CHIN, K. O. A.; MORROW, J. R.; LAKE, C. H.; CHURCHILL, M. R., *Inorg. Chem.* **1994**, *33*, 656–664.

[6] SASAI, H.; SUZUKI, T.; ITOH, N.; TANAKA, K.; DATE, T.; OHAMURA, K.; SHIBASAKI, M., *J. Am. Chem. Soc.* **1993**, *115*, 10372–10373.

[7] SASAI, H.; SUZUKI, T.; ITOH, N.; ARAI, S.; SHIBASAKI, M., *Tetrahedron Lett.* **1993**, *34*, 2657–2660.

[8] GRÖGER, H.; SAIDA, Y.; SASAI, H.; YAMAGUCHI, K.; MARTENS, J.; SHIBASAKI, M., *J. Am. Chem. Soc.* **1998**, *120*, 3089–3103.

[9] SALVADORI, P.; SUPERCHI, S.; MINUTOLO, F., *J. Org. Chem.* **1996**, *61*, 4190–4191.

[10] KOLB, H. C.; VAN NIEUWEHNZE, M. S.; SHARPLESS, K. B., *Chem. Rev.* **1994**, *94*, 2483–2547.

[11] MIAO, G.; ROSSITER, B. E., *J. Org. Chem.* **1995**, *60*, 8424–8427.

[12] KAWASAKI, K.; KATSUKI, T., *Tetrahedron* **1997**, *53*, 6337–6350.

[13] MACURA, S.; WÜTHRICH, K.; ERNST, R. R., *J. Magn. Reson.* **1982**, *47*, 351–357.

[14] MARION, D.; WÜTHRICH, K., *Biochem. Biophys. Res. Commun.* **1983**, *113*, 967–974.

[15] BAX, A.; DAVIS, D. G., *J. Magn. Reson.* **1985**, *65*, 355-360.

[16] DEROME, A.; WILLIAMSON, M., *J. Magn. Reson.* **1990**, *88*, 177–185.

[17] CAVANAGH, J.; FAIRBROTHER, W. J.; PALMER, III, A. G.; SKELTON, N. J., *Protein NMR spectroscopy - Principles and practice;* ACADEMIC PRESS: SAN DIEGO, 1996.

[18] CASTIGLIONI, E., "NIR OPERATION WITH CONVENTIONAL CD SPECTROPOLARIMETERS", IN *Book of Abstracts, 6th International Conference on CD.* PISA, 1997 .

[19] BEEBY, A.; CLARKSON, I. M.; DICKINS, R. S.; FAULKNER, S.; PARKER, D.; ROYLE, L.; DE SOUSA, A. S.; WILLIAMS, J. A. G., *J. Chem. Soc., Perkin Trans. 2* **1999**, 493–503.

[20] BARRY, C. D.; NORTH, A. C. T.; GLASEL, J. A.; WILLIAMS, R. J. P.; XAVIER, A. V., *Nature* **1971**, *232*, 236–245.

[21] PETERS, J. A.; HUSKENS, J.; RABER, D. J., *Progr. NMR Spectrosc.* **1996**, *28*, 283–350.

[22] KERÉNYI, K., *Die Mythologie der Griechen;* ITALIAN TRANSLATION, GARZANTI: MILANO, 5TH ED. ED.; 1989.

Appendix B

List of numbered molecules

DOTA **1**

(R)-DOTMA
(R)-**2**

(R)-DOPEA
(R)-**3**

(S)-DONEA
(S)-**4**

(S)-DOTAM-Ile
(S)-**5**

(S)-DOTAM-Phe
(S)-6

(S)-DOTAM-Pro
(S)-7

(S)-THP
(S)-8

YbNa₃BINOL₃
(S)-9

LuNa₃BINOL₃
(S)-10

YbK₃BINOL₃
(S)-11

Yb(fod)₃ 12

13

14

15

16

17

18

19

20

21

22

23

24

Daunorubicin (**25**)

MEN 10755 (**26**)

Appendix C

Names and abbreviations

acac	acethylacetonate
AcO	acetate
BINOL	2-binaphthol
BSA	bovine serum albumine
CD	circular dichroism
CD	cyclodextrin (in chapter 8)
CFP	crystal field parameter
CID	collision induced dissociation
COSY	correlation spectroscopy
CP	coordination polyhedron
Dau	(8S-cis)-8-acetyl-10-[(3-amino-2,3,6-trideoxy-α-L-lyxo-hexa-pyranosyl)oxy]-7,8,9,10-tetrahydro-6,8,11-tri-hydroxy-l-methoxy-5,12-naphthacenedione (daunomycin, **26**)
DMF	dimethyl formamide
DMSO	dimethyl sulfoxide
DO3MA	(1R,4R,7R)-α,α',α'' - trimethyl-1,4,7-triazacyclododecane triacetic acid
DOPEA	1,4,7,10-tetrakis[(R)-1-(phenyl)ethylcarbamoylmethyl]-1,4,7,10-tetraazacyclododecane, (R)-**3**
DONEA	1,4,7,10-tetrakis[(R)-1-(1-naphtyl)ethylcarbamoylmethyl]-1,4,7,10-tetraazacyclododecane, (R)-**4**
DOTA	1,4,7,10-tetraazacyclododecane tetraacetic acid, **1**
DOTAM	1,4,7,10-tetrakis(2-carbamoilethyl)-1,4,7,10-tetraazaciclo-dodecane

DOTAM-Ile	1,4,7,10-tetrakis[(S)-1-(carboxymethyl)-2-(methyl)-propyl-carbamoylmethyl]-1,4,7,10-tetraazacyclododecane, (S)-**5**
DOTAM-Phe	1,4,7,10-tetrakis[(S)-1-(carboxymethyl)-2-(phenyl)-ethyl-carbamoylmethyl]-1,4,7,10-tetraazacyclododecane, (S)-**6**
DOTAM-Pro	1,4,7,10-tetrakis[(S)-2-(carboxymethyl)-pyrrolidinylamoyl-methyl]-1,4,7,10-tetraazacyclododecane, (S)-**7**
DOTMA	(1R,4R,7R,10R) -$\alpha,\alpha',\alpha'',\alpha'''$ - tetramethyl-1,4,7,10-tetra-azacyclododecane tetraacetic acid, (R)-**2**
DOTPBz$_4$	1,4,7,10-tetraazacyclododecane tetrakis(methylenebenzyl-phosphinic acid)
dpm	dipivalomethanate
DTPA	diethylentriamino-N,N,N',N'',N''-pentaacetic acid
EDTA	1,2-ethylendiamino-N,N,N',N'-tetraacetic acid
ESI-MS	electrospray ionization mass spectroscopy
FID	free induction decay
fod	6,6,7,7,8,8,8-heptafluoro-2,2-dimethyl-3,5-octanedioate
HETCOR	heteronuclear correlation
KHMDS	potassium hexamethyl disilazane or potassium bis(trimethylsilil) amide, $KN[Si(CH_3)_3]_2$
LIR	lanthanide induced relaxation
LIS	lanthanide induced shift
MEN 10755	7-O-[2,6-dideoxy-4-O-(2,3,6-trideoxy-3-amino-α-L-lyxo-hexopyranosyl)]-4-demethoxy-14-hydroxydaunomycinone, **26**
MRI	magnetic resonance imaging
NIR	near infra-red
NOE	nuclear overhauser effect
NOESY	nuclear overhauser effect spectroscopy
NMR	nuclear magnetic resonance
OR	orifice voltage
PRE	paramagnetic relaxation enhancement
SAP	square antiprism

TETA   1,4,8,11-tetraazaciclotetradecane-N,N',N'',N'''-tetraacetic acid

TfO   trifluoromethane sulphonate (triflate)

THED   1,4,7,10-tetrakis(2-hydroxyethyl)-1,4,7,10-tetraazacyclodo-decane

THF   tetrahydrofuran

THP   1,4,7,10-tetrakis[(S)-2-hydroxypropyl]-1,4,7,10-tetraaza-cyclododecane, (S)-**8**

TOCSY   total correlation spectroscopy

TSA   twisted square antiprism

UV   ultra-violet

Vis   visible

Elenco delle Tesi di perfezionamento della Classe di Scienze
pubblicate dall'Anno Accademico 1992/93

HISAO FUJITA YASHIMA, *Equations de Navier-Stokes stochastiques non homogènes et applications*, 1992.

GIORGIO GAMBERINI, *The minimal supersymmetric standard model and its phenomenological implications*, 1993.

CHIARA DE FABRITIIS, *Actions of Holomorphic Maps on Spaces of Holomorphic Functions*, 1994.

CARLO PETRONIO, *Standard Spines and 3-Manifolds*, 1995.

MARCO MANETTI, *Degenerations of Algebraic Surfaces and Applications to Moduli Problems*, 1995.

ILARIA DAMIANI, *Untwisted Affine Quantum Algebras: the Highest Coefficient of* det H_η *and the Center at Odd Roots of 1*, 1995.

FABRIZIO CEI, *Search for Neutrinos from Stellar Gravitational Collapse with the MACRO Experiment at Gran Sasso*, 1995.

ALEXANDRE SHLAPUNOV, *Green's Integrals and Their Applications to Elliptic Systems*, 1996.

ROBERTO TAURASO, *Periodic Points for Expanding Maps and for Their Extensions*, 1996.

YURI BOZZI, *A study on the activity-dependent expression of neurotrophic factors in the rat visual system*, 1997.

MARIA LUISA CHIOFALO, *Screening effects in bipolaron theory and high-temperature superconductivity*, 1997.

DOMENICO M. CARLUCCI, *On Spin Glass Theory Beyond Mean Field*, 1998.

RENATA SCOGNAMILLO, *Principal G-bundles and abelian varieties: the Hitchin system*, 1998.

GIACOMO LENZI, *The MU-calculus and the Hierarchy Problem*, 1998.

GIORGIO ASCOLI, *Biochemical and spectroscopic characterization of CP20, a protein involved in synaptic plasticity mechanism*, 1998.

FABIO PISTOLESI, *Evolution from BCS Superconductivity to Bose-Einstein Condensation and Infrared Behavior of the Bosonic Limit*, 1998.

LUIGI PILO, *Chern-Simons Field Theory and Invariants of 3-Manifolds*, 1999.

PAOLO ASCHIERI, *On the Geometry of Inhomogeneous Quantum Groups*, 1999.

SERGIO CONTI, *Ground state properties and excitation spectrum of correlated electron systems*, 1999.

GIOVANNI GAIFFI, *De Concini-Procesi models of arrangements and symmetric group actions*, 1999.

DONATO NICOLÒ, *Search for neutrino oscillations in a long baseline experiment at the Chooz nuclear reactors*, 1999.

ROCCO CHIRIVÌ, *LS algebras and Schubert varieties*, 2000.

FRANCESCO MATTIA ROSSI, *A Study on Nerve Growth Factor (NGF) Receptor Expression in the Rat Visual Cortex: Possible Sites and Mechanisms of NGF Action in Cortical Plasticity*, 2000.

VALENTINO MAGNANI, *Elements of Geometric Measure Theory on Sub-Riemannian Groups*, 2002.

GUIDO PINTACUDA, *NMR and NIR-CD of Lanthanide Complexes*, 2004.

"CompoMat" Loc. Braccone, 02040 Configni (RI), Italy
Finito di stampare nel mese di dicembre del 2004